Pediatric Oncology

A. Lindsay Frazier • James F. Amatruda

Editors

Pediatric Germ Cell Tumors

Biology Treatment Survivorship

 Springer

Editors
A. Lindsay Frazier
Dana-Farber Cancer Institute
Pediatric Oncology
Boston, MA
USA

James F. Amatruda
UT Southewestern Medical Center
Department of Pediatrics, Molecular
Biology and Internal Medicine
Dallas, TX
USA

ISBN 978-3-642-38970-2   ISBN 978-3-642-38971-9   (eBook)
DOI 10.1007/978-3-642-38971-9
Springer Heidelberg New York Dordrecht London

Acknowledgements and Dedication

We would like first of all to acknowledge the authors for their hard work and dedication to producing informative, readable and up-to-date chapters for this volume. We have enjoyed the many discussions and interactions with this wonderful group of people. We also thank our many colleagues in the Children's Oncology Group, at the Dana-Farber Cancer Institute and UT Southwestern Medical Center for their support and for the excellent care they provide for children with germ cell tumors and other cancers. Thanks also to the production and editorial staff at Springer DE, particularly Ms. Meike Stoeck who has patiently shepherded us through the editorial process. We particularly wish to thank our families – Ted Amatruda, Lisa Pon, Owen, Aidan, Andre and Nate Dempsey – for their unwavering patience and loving support.

Finally, we would like to recognize our dear friend, colleague and mentor, Carlos Rodriguez-Galindo. A friend to all and a giant in the field who wears his many accomplishments lightly, no one has been a more effective champion for children with germ cell tumor and other rare cancers. It is in recognition of Carlos's inspiring leadership and many kindnesses that we gratefully dedicate this volume to him.

Contents

Introduction

The clinical management of pediatric germ cell tumors (GCTs) carries with it a unique set of challenges. Incidence data show two distinct peaks in GCT incidence, one in young children ages 0–4 and another extending from the onset of puberty through young adulthood. The incidence of GCT among adolescents and young adults is rapidly increasing, for unclear reasons (Frazier and Amatruda 2009). Although germ cell tumors are often classified as "rare" cancers, GCTs in fact account for 15 % of the malignancies diagnosed during adolescence, and testicular GCTs are the most common cancer in young men aged 15–40 (Oosterhuis and Looijenga 2005). The broad age distribution of GCTs, from infancy to the mid-40s, means that patients with GCTs may be cared for by pediatric oncologists, medical oncologists, radiation oncologists, general surgeons, urologists and/or gynecologic oncologists. This wide range of clinical settings has led to the adoptions of a variety of different management strategies for GCTs. Complicating matters is the fact that GCTs, which arise from pluripotent primordial germ cells, present in a wide range of subtypes with different histologies, marker expression and clinical behavior. Moreover, germ cells develop in the context of somatic cells of the gonad, which can themselves give rise to a unique set of tumors with their own biological and clinical features.

GCTs became curable with the advent of cisplatin chemotherapy nearly 40 years ago (Einhorn and Donohue 1977; Williams et al. 1987). Since that time, a series of careful studies has established cisplatin-based multiagent regimens that have proved highly effective even in the setting of advanced disease. While these regimens remain the mainstay of GCT therapy, it is important to note several ongoing challenges, the first of which is that cisplatin-based therapy persistently fails to cure about 15 % of patients (Einhorn 2002). Survival is especially poor for patients with mediastinal GCTs, those with metastatic disease at presentation, or those whose tumors are resistant to cisplatin (Frazier and Amatruda 2009).

Secondly, there is significant therapy-related toxicity in patients treated for GCTs. Late effects in patients treated with cisplatin, etoposide and bleomycin (the current first-line regimen for GCT) include pulmonary fibrosis (Osanto et al. 1992), renal insufficiency and salt-wasting (Bosl et al. 1986; Hansen et al. 1988; Bokemeyer et al. 1996), infertility and hormonal changes

(Hansen et al. 1990; Hansen and Hansen 1993; Berger et al. 1996; Strumberg et al. 2002; Huddart et al. 2003), hyperlipidemia (Bissett et al. 1990; Gietema et al. 1992), Raynaud's phenomenon (Teutsch et al. 1977; Vogelzang et al. 1981; Berger et al. 1996; Bokemeyer et al. 1996), obesity (Bissett et al. 1990; Boyer et al. 1990; Gietema et al. 1992; Siviero-Miachon et al. 2008, 2009) and neuropathy (Glendenning et al. 2010). Ototoxicity results in significant hearing loss, as assessed by audiogram, in nearly 80 % of patients (Bokemeyer et al. 1996; Strumberg et al. 2002). Most troubling, emerging evidence indicates a doubling of the risk of early onset cardiovascular disease (Huddart et al. 2003) and second malignancies (Travis et al. 1997, 2005) in survivors of germ cell tumor treatment. There is evidence that children are particularly vulnerable to late effects of therapy, especially ototoxicity and pulmonary abnormalities (Hale et al. 1999). In a large cohort study, the cumulative risk of secondary malignancy also increased with decreasing age at diagnosis (Travis et al. 2005).

Finally, it is important to note that the majority of the clinical trial experience in GCT has focused on adult men with testicular (T)GCT, who represent the largest patient population. While regimens for adult TGCT have been widely adopted for other patient populations, it is less clear that all of the lessons from these trials, and the therapies that they have advanced, apply equally to other groups such as children with gonadal or extragonadal GCT, and women with ovarian GCT.

Both childhood and adolescent/adult GCTs are thought to originate from primordial germ cells (PGCs), pluripotent cells that can take on a variety of different histologic fates. Thus GCTs can be germinomas (which retain primitive germ cell characteristics); tumors of extraembryonic tissues (yolk sac, trophoblast); and tumors differentiated to endoderm, mesoderm, and ectoderm (teratomas). Type I GCTs occur in infants and children and consist of benign teratomas and malignant yolk sac tumors (YSTs). Type II tumors in adolescents and adults have more diverse histology, and include germinomas (seminomas in males, dysgerminomas in females), embryonal carcinoma, teratoma, yolk sac tumor and choriocarcinoma. Based on differing epidemiology, clinical outcome and histologic spectrum, it has long been suspected that GCTs in young children may be biologically distinct from GCTs occurring in older (post-pubertal) populations.

This idea is increasingly supported by molecular evidence from studies comparing genome structure and gene expression in GCTs from different patient populations. For example, Type I tumors show variable loss of imprinting (LOI) at loci such as IGF2 and H19 (Ross et al. 1999; Schneider et al. 2001), and cytogenetic data consistently show loss of chromosome 1p and 6q. In contrast, type II tumors show complete LOI, implying they originate from a different stage of embryonic germ cell than do the type I tumors. Type II tumors also commonly exhibit amplification of chromosome 12p. More recently, studies directly comparing the gene expression patterns of pediatric and adult GCTs have demonstrated distinct transcriptional profiles in tumors of similar histology arising in different age groups (for example, in yolk sac tumors of children vs. yolk sac tumors of adolescent/adults) (Palmer et al. 2008). Taken together, these studies support the notion that different biological mechanisms may drive childhood and adolescent/adult germ cell tumorigenesis.

While progress is being made, there is still much to be done. In some ways the very success of cisplatin-based therapy has reduced pressure on the field to investigate alternative strategies. A major challenge going forward will be to translate insights from clinical and molecular studies into new approaches that will improve patient outcomes. For example, it will be critical to define robust biomarkers that predict risk of disease progression and treatment resistance. Such a biomarker could allow much more accurate risk stratification and tailoring of therapy, sparing many patients the deleterious side effects of current GCT therapies. As another example, high-resolution molecular profiling of cisplatin-resistant GCTs could potentially be used to guide therapy with more specific, targeted agents. Advances like these will require ongoing collaborations between basic scientists, clinical researchers, clinicians, patients and patient advocates.

In this volume, we present practical management advice for clinicians caring for children with GCTs, along with a comprehensive review of GCT pathology. Specific attention is given to the unique features of and approaches required for ovarian GCTs and gonadal-stromal tumors. We also provide updates on the latest research on GCT epidemiology, biology and late effects. It is our hope that this book will serve as both a practical guide and a stimulus for ongoing research and collaboration, for the benefit of all our patients.

A. Lindsay Frazier, MD, ScM
James F. Amatruda, MD, PhD

References

Berger CC, Bokemeyer C, Schuppert F, Schmoll HJ (1996) Endocrinological late effects after chemotherapy for testicular cancer. Br J Cancer 73:1108–1114

Bissett D, Kunkeler L, Zwanenburg L, Paul J, Gray C, Swan IR et al (1990) Long-term sequelae of treatment for testicular germ cell tumours. Br J Cancer 62:655–659

Bokemeyer C, Berger CC, Kuczyk MA, Schmoll HJ (1996) Evaluation of long-term toxicity after chemotherapy for testicular cancer. J Clin Oncol 14:2923–2932

Bosl GJ, Leitner SP, Atlas SA, Sealey JE, Preibisz JJ, Scheiner E (1986) Increased plasma renin and aldosterone in patients treated with cisplatin-based chemotherapy for metastatic germ-cell tumors. J Clin Oncol 4:1684–1689

Boyer M, Raghavan D, Harris PJ, Lietch J, Bleasel A, Walsh JC et al (1990) Lack of late toxicity in patients treated with cisplatin-containing combination chemotherapy for metastatic testicular cancer. J Clin Oncol 8:21–26

Einhorn LH (2002) Chemotherapeutic and surgical strategies for germ cell tumors. Chest Surg Clin N Am 12:695–706

Einhorn LH, Donohue JP (1977) Improved chemotherapy in disseminated testicular cancer. J Urol 117:65–69

Frazier AL, Amatruda JF (2009) Germ cell tumors. In: Orkin SH, Fisher DE, Look AT, et al. (eds). Oncology of infancy and childhood, 7th edition. Philadelphia, Saunders Elsevier, pp 911–961

Gietema JA, Sleijfer DT, Willemse PH, Schraffordt Koops H, van Ittersum E, Verschuren WM et al (1992) Long-term follow-up of cardiovascular risk factors in patients given chemotherapy for disseminated nonseminomatous testicular cancer. Ann Intern Med 116:709–715

Glendenning JL, Barbachano Y, Norman AR, Dearnaley DP, Horwich A, Huddart RA (2010) Long-term neurologic and peripheral vascular toxicity after chemotherapy treatment of testicular cancer. Cancer 116:2322–2331

Hale GA, Marina NM, Jones-Wallace D, Greenwald CA, Jenkins JJ, Rao BN et al (1999) Late effects of treatment for germ cell tumors during childhood and adolescence. J Pediatr Hematol Oncol 21:115–122

Hansen PV, Hansen SW (1993) Gonadal function in men with testicular germ cell cancer: the influence of cisplatin-based chemotherapy. Eur Urol 23:153–156

Hansen SW, Groth S, Daugaard G, Rossing N, Rorth M (1988) Long-term effects on renal function and blood pressure of treatment with cisplatin, vinblastine, and bleomycin in patients with germ cell cancer. J Clin Oncol 6:1728–1731

Hansen SW, Berthelsen JG, von der Maase H (1990) Long-term fertility and Leydig cell function in patients treated for germ cell cancer with cisplatin, vinblastine, and bleomycin versus surveillance. J Clin Oncol 8:1695–1698

Huddart RA, Norman A, Shahidi M, Horwich A, Coward D, Nicholls J et al (2003) Cardiovascular disease as a long-term complication of treatment for testicular cancer. J Clin Oncol 21:1513–1523

Oosterhuis JW, Looijenga LH (2005) Testicular germ-cell tumours in a broader perspective. Nat Rev Cancer 5:210–222

Osanto S, Bukman A, Van Hoek F, Sterk PJ, De Laat JA, Hermans J (1992) Long-term effects of chemotherapy in patients with testicular cancer. J Clin Oncol 10:574–579

Palmer RD, Barbosa-Morais NL, Gooding EL, Muralidhar B, Thornton CM, Pett MR et al (2008) Pediatric malignant germ cell tumors show characteristic transcriptome profiles. Cancer Res 68:4239–4247

Ross JA, Schmidt PT, Perentesis JP, Davies SM (1999) Genomic imprinting of H19 and insulin-like growth factor-2 in pediatric germ cell tumors. Cancer 85:1389–1394

Schneider DT, Schuster AE, Fritsch MK, Hu J, Olson T, Lauer S et al (2001) Multipoint imprinting analysis indicates a common precursor cell for gonadal and nongonadal pediatric germ cell tumors. Cancer Res 61:7268–7276

Siviero-Miachon AA, Spinola-Castro AM, Guerra-Junior G (2008) Detection of metabolic syndrome features among childhood cancer survivors: a target to prevent disease. Vasc Health Risk Manag 4:825–836

Siviero-Miachon AA, Spinola-Castro AM, Guerra-Junior G (2009) Adiposity in childhood cancer survivors: insights into obesity physiopathology. Arq Bras Endocrinol Metabol 53:190–200

Strumberg D, Brugge S, Korn MW, Koeppen S, Ranft J, Scheiber G et al (2002) Evaluation of long-term toxicity in patients after cisplatin-based chemotherapy for non-seminomatous testicular cancer. Ann Oncol 13:229–236

Teutsch C, Lipton A, Harvey HA (1977) Raynaud's phenomenon as a side effect of chemotherapy with vinblastine and bleomycin for testicular carcinoma. Cancer Treat Rep 61:925–926

Travis LB, Curtis RE, Storm H, Hall P, Holowaty E, Van Leeuwen FE et al (1997) Risk of second malignant neoplasms among long-term survivors of testicular cancer. J Natl Cancer Inst 89:1429–1439

Travis LB, Fossa SD, Schonfeld SJ, McMaster ML, Lynch CF, Storm H et al (2005) Second cancers among 40,576 testicular cancer patients: focus on long-term survivors. J Natl Cancer Inst 97:1354–1365

Vogelzang NJ, Bosl GJ, Johnson K, Kennedy BJ (1981) Raynaud's phenomenon: a common toxicity after combination chemotherapy for testicular cancer. Ann Intern Med 95:288–292

Williams SD, Birch R, Einhorn LH, Irwin L, Greco FA, Loehrer PJ (1987) Treatment of disseminated germ-cell tumors with cisplatin, bleomycin, and either vinblastine or etoposide. N Engl J Med 316:1435–1440

Biology of Germ Cell Tumors

1

Matthew Jonathan Murray and Stefan Schönberger

Contents

M.J. Murray (✉)
Department of Pathology, University of Cambridge, Cambridge, UK

Department of Paediatric Haematology and Oncology, Cambridge University Hospitals NHS Foundation Trust, Cambridge, UK
e-mail: mjm16@cam.ac.uk

S. Schönberger, MD
Department of Pediatric Hematology and Oncology, University of Bonn,
University Children's Hospital Bonn, Bonn, Germany
e-mail: stefan.schoenberger@ukb.uni-bonn.de

1.1 Introduction

An improved understanding of the biology of germ cell tumors (GCTs) is essential for two main reasons. Firstly, biology will be useful in identifying patients with good clinical outcomes, who may in the future be able to safely receive less therapy and hence experience fewer late effects of treatment, without compromising excellent outcome (Mann et al. 2000; Göbel et al. 2002). Secondly, for some groups of patients, such as older pediatric patients (greater than 11 years of age) with advanced (stage 4) extragonadal disease, outcomes remain markedly inferior (Marina et al. 2006). Discovering a prognostic gene "signature" that may be used for risk stratification in such patients will be important. This process will also highlight the genes and pathways critical in the pathogenesis of poor-risk tumors and therefore represents an important step towards identifying targets for the development of novel therapeutic agents, with favorable toxicity profiles, which may be used in these subgroups.

Most research in GCTs has been performed on adult testicular cases (TGCTs), which have the highest incidence of all GCTs (Murray et al. 2009). However, a number of factors have historically limited the ability to study pediatric GCT biology. As GCTs occur throughout childhood, from the neonatal period into adulthood (Schneider et al. 2004; Murray and Nicholson 2010), patients with these tumors present to a

A.L. Frazier, J.F. Amatruda (eds.), *Pediatric Germ Cell Tumors*, Pediatric Oncology 1,
DOI 10.1007/978-3-642-38971-9_1, © Springer-Verlag Berlin Heidelberg 2014

wide variety of medical practitioners, including neonatologists, general pediatricians, family practitioners, urologists, general and gynecologic surgeons and, for those with intracranial disease, neurosurgeons. Collection of appropriate tumor specimens has therefore historically been piece-meal and limited to individual, large treatment centers (Kaatsch 2004). Furthermore, GCTs are relatively rare, comprising only approximately 2–4 % of childhood cancers (Schneider et al. 2004; Murray and Nicholson 2010), and conse-quently, most early reports studied very small numbers of patient cases. Moreover, GCTs are highly heterogeneous, characterized by variable histologic subtypes and anatomic sites of disease (gonadal and extragonadal), limiting the numbers of specimens studied in each subtype and site. These limitations have more recently been over-come by the establishment of tumor banks, such as that set up in the UK by the Children's Cancer and Leukaemia Group (CCLG), in Germany by the MAKEI Study Group, and in the USA by the Children's Oncology Group (COG). Additionally, international collaboration between national chil-dren's oncology groups has increased the num-bers of specimens available for analysis.

1.2   Embryology and Development

GCTs are all believed to arise from a common progenitor cell, the primordial germ cell (PGC) (Teilum 1965) (Fig. 1.1). These cells represent the embryonic precursors of gametes, namely spermatogonia in the male and oocytes in the female. PGCs are totipotent, meaning they have the ability to differentiate into all three embry-ologic layers (endoderm, ectoderm, and meso-derm) in addition to extraembryonic tissues [namely yolk sac and trophoblastic (placenta-like) differentiation]. In early human devel-opment, PGCs separate from the epiblast and migrate within the wall of an outpouching of the extraembryonic yolk sac at approximately 3 weeks gestation (Nicholson and Palmer 2010). They subsequently migrate along the vitelline duct, the wall of the hindgut and the dorsal mesentery to the genital ridge. This process is completed between 6 and 7 weeks of gesta-tion and is dependent upon the strict temporo-spatial expression of a number of additional mediators, including the chemokine CXCL12 (SDF1) and its receptor CXCR4 (Doitsidou et al. 2002; Molyneaux et al. 2003). CXCL12 is known to be regulated by short, nonprotein-coding RNAs termed microRNAs (Staton et al. 2011). Importantly, both CXCL12 and CXCR4 have been shown to play important roles in the development of GCTs. Aberrant expression of CXCL12 induces disordered migration of germ cells (Molyneaux et al. 2003), and this is believed to account for the extragonadal sites at which GCTs may occur. Consistent with this observation, extragonadal GCTs have been shown to express CXCR4, coincident with areas of known CXCL12 expression in utero (Gilbert et al. 2009). In addition, the *KIT ligand* gene (*KITLG*, also known as *steel factor*) is expressed in an increasing gradient from the yolk sac to the gonadal ridge and influences the motility, rather than the directionality, of PGCs (Gu et al. 2009, 2011). Those PGCs that migrate appropriately to the gonadal ridge subsequently undergo gender-specific differ-entiation into oocytes or spermatogonia; those that do not usually undergo BAX-dependent apoptosis (Stanley et al. 2007; Gu et al. 2011). Consequently, mutation of *BAX* and loss of BAX expression was shown to occur in a subset of extragonadal pediatric GCTs (Addeo et al. 2007). In addition to aberrant PGC migration, epigenetic (see Sect. 1.3) and environmen-tal factors may contribute further towards the development of GCTs and the observation that some histologic GCT subtypes are predomi-nantly restricted to specific anatomic sites.

Recent advances in the understanding of stem cell and germ cell biology have allowed the demonstration of similarities between PGCs and embryonic stem cells (ESCs). Both cell types actively express genes known to regulate pluripotency during embryogenesis such as *POU5F1 (OCT3/4), NANOG, SOX2*

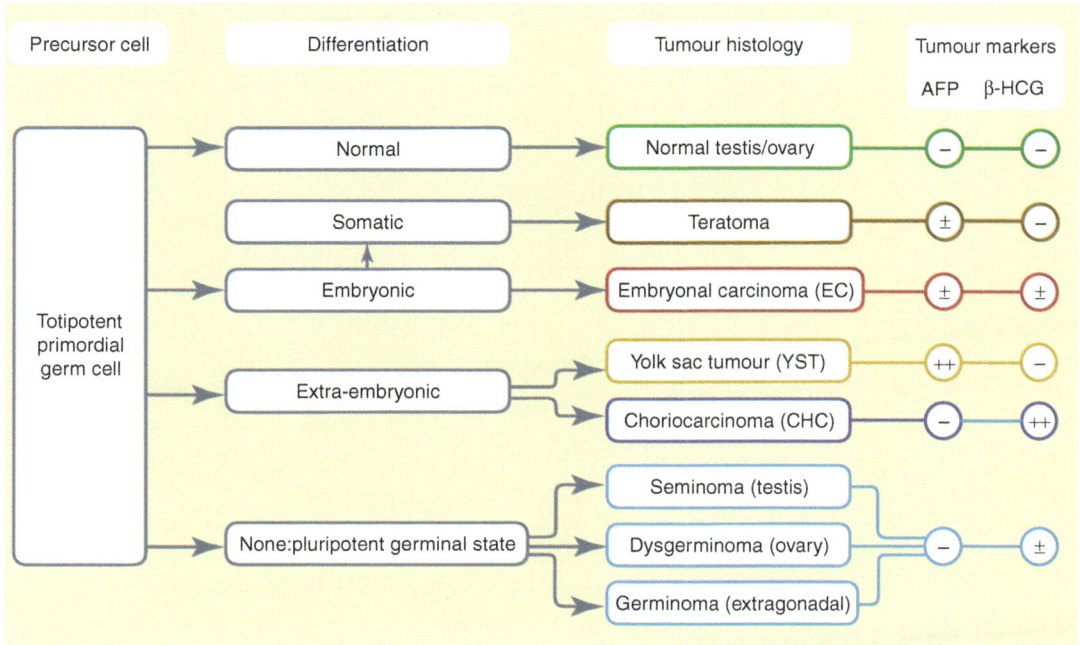

Fig. 1.1 Derivation of germ cell tumors from primordial germ cells (Reprinted with permission from Murray and Nicholson (2010))

and *LIN28*. Thus, differentiation of murine ESCs into putative PGCs is possible, dependent on the expression of *LIN28* and/or BLIMP1 (West et al. 2009). In contrast to other pluripotency markers such as *POU5F1* (*OCT3/4*) and *NANOG*, immunohistochemical analysis has demonstrated that *LIN28* is detectable in prespermatogonia (Gillis et al. 2011) and acts as a marker for GCTs regardless of their anatomic localization, patient age, or histologic subtype (Cao et al. 2011a, b; Gillis et al. 2011; Xue et al. 2011). *LIN28* maintains the undifferentiated state of subsets of malignant GCTs (Gillis et al. 2011) and is also known to be a negative regulator of the *let-7* microRNA family (Viswanathan et al. 2008, 2009). Recently, through RNA interference mediated *LIN28* depletion, it has been shown that *LIN28* expression in malignant GCTs directly results in *let-7* downregulation, causing a significant concomitant upregulation of important cancer-associated protein-coding genes (Murray et al. 2013). Such studies dem-

onstrate that the *LIN28/let-7* axis in malignant GCTs is a potential target for the development of novel therapeutic agents (Murray et al. 2013).

Similarly to *LIN28*, the *epithelial cell adhesion molecule* (*EPCAM*) gene is increasingly expressed in primary teratomas with loss of differentiation and also in malignant GCTs, irrespective of age, sex, site and clinical stage of the tumor (Schönberger et al. 2013) (Fig. 1.2). Of note, EPCAM is highly expressed on the surface of undifferentiated human ESCs and lost during ESC differentiation (Lu et al. 2010; Ng et al. 2010). Consequently, EPCAM has been used to identify circulating tumor cells in ovarian cancer, the presence of which are associated with poor prognosis (Poveda et al. 2011). Analysis for EPCAM-positive cells in peripheral blood, and correlation with the presence of metastatic disease and patient outcome, may therefore be of future interest in patients with GCTs.

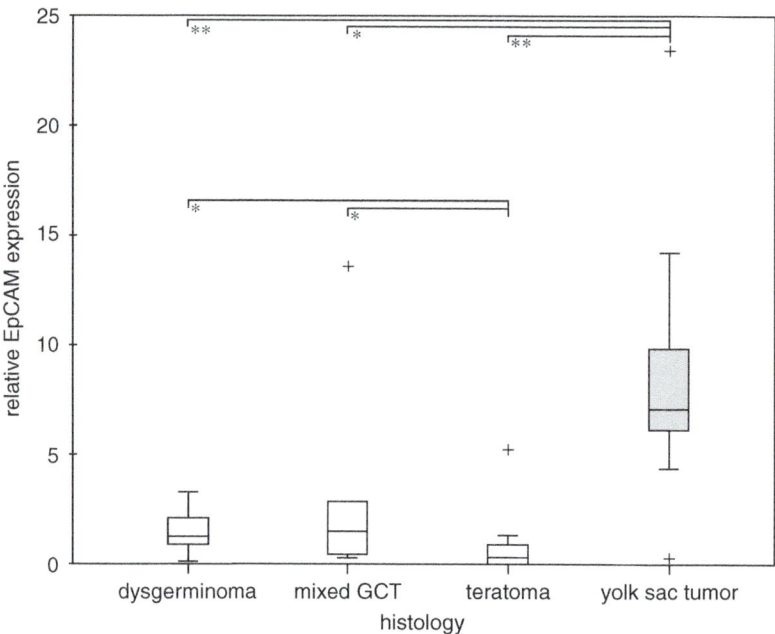

Fig. 1.2 Expression of *EPCAM* in germ cell tumors. Real-time quantitative PCR analysis of *EPCAM* mRNA expression in 16 mature and immature teratomas, 6 mixed malignant non-seminomatous germ cell tumors (*GCT*), 19 yolk sac tumors, and 7 dysgerminomas. Relative *EPCAM* expression levels were normalized against *POLR2A* and *GUSB* and presented as $2^{-\Delta Ct}$ values. **P-value <0.0001, *P-value <0.05 (From EpCAM – a novel molecular target for the treatment of pediatric and adult germ cell tumors. Schönberger et al. (2013). Copyright © 2012 Wiley Periodicals, Inc.)

1.3 Epigenetics: Genomic Imprinting and Methylation

1.3.1 Studies Which Demonstrate That Epigenetic Changes in GCTs Reflect the Stage of Development of the Primordial Germ Cell at the Time of Malignant Transformation

Differentiation of PGCs into gametes during normal human development is under the control of epigenetic mechanisms. Studying such phenomena provides insights into the stage at which aberrant PGCs transform into GCTs. Initially, PGCs are characterized by erasure of imprinting and thus bi-allelic gene expression. Subsequently, during PGC development into mature gametes, they undergo genomic imprinting (Schneider et al.

2001a). As a result, expression of these genes is only from a single maternal or paternal allele in mature gametes. Imprinted genes such as *SNRPN* (normally paternally expressed in somatic tissues), *H19* (maternally expressed), and *IGF2* (paternally expressed) are therefore suitable candidates to investigate the methylation status and allele-specific gene expression of GCTs. Studies have shown that both gonadal and non-gonadal pediatric GCTs are derived from PGCs that have consistently lost the imprinting of *SNRPN* and partial imprinting loss of *H19* and *IGF2*. This observation likely reflects the preservation of the physiologic erasure of the *IGF2/H19* methylation imprint in PGCs, rather than a loss of imprinting characteristic of somatic tumors (Sievers et al. 2005). A study of *SNRPN* methylation in both pediatric and adult GCTs revealed that the majority had a non-somatic (hypo) methylation pattern, reflecting that of the PGC giving rise to the tumor (Bussey et al. 2001). The above studies are consistent with the theory that all malignant GCTs, both gonadal and extragonadal, arise from a common origin, the PGC. However, an alternative theory has been suggested for intracranial tumors, whereby neural stem cells may represent the more likely cells of origin for GCTs at this site (rather than PGCs that have migrated aberrantly to the central nervous system), as these endogenous progenitor cells also demonstrate a lack of *SNRPN* methylation, as observed in PGCs (Lee et al. 2011).

1.3.2 Studies Which Demonstrate That Epigenetic Changes in GCT Subtypes May Account for the Observed Differences in Clinical Behavior and Outcome

Although overall treatment outcomes for patients with malignant GCTs are excellent, differences do exist by histologic subtype. For example, 5-year overall survival is well in excess of 90 % for intracranial germinomas treated with radiotherapy alone (Bamberg et al. 1999) but less than 50 % for non-germinomatous (e.g., yolk sac) tumors treated in an identical fashion, with many early relapses (Hoffman et al. 1991). For the latter group, 5-year relapse-free survival only increases to 67 % with the addition of systemic chemotherapy (Calaminus et al. 2005). Similarly, in both pediatric and adult patients, outcomes in extracranial non-germinomatous tumors are inferior to germinomas (Marina et al. 2006; Stang et al. 2012). Biologic studies investigating methylation status in pediatric malignant GCTs have attempted to explain these differences in outcome. Global methylation array analysis of gene regulatory regions in malignant GCTs revealed that yolk sac tumors (YSTs) have increased methylation at many loci, including silencing of potential tumor-suppressor genes and those associated with apoptosis (Jeyapalan et al. 2011). Furthermore, this was associated with higher levels of expression of the methyltransferase gene *DNMT3b* in YSTs, suggesting a mechanism underlying the phe-

notype and a potential explanation for the more aggressive natural history displayed by YSTs compared with germinomas (Jeyapalan et al. 2011). With regard to other specific genes, studies have shown that the tumor-suppressor genes *APC* (Kato et al. 2006; Schönberger et al. 2010; Jeyapalan et al. 2011; Okpanyi et al. 2011) and *SFRP2* (Schönberger et al. 2010) are methylated in pediatric YSTs but not in germinomas, resulting in activation of the *Wnt* signaling pathway (Fritsch et al. 2006; Palmer et al. 2008) (see also Sect. 1.6).

In adult TGCTs, methylation of *APC*, *RASSF1A*, and *MGMT* was shown to occur more frequently in non-seminomatous tumors compared with seminomas (Honorio et al. 2003). Subsequently, methylation of the tumor-suppressor genes *RASSF1A* and *HIC1* have been shown to be associated with treatment resistance, while methylation of *MGMT* was associated with platinum sensitivity (Koul et al. 2004). Similarly, in methylation studies of pediatric tumors, *APC*, *RASSF1A*, and *HIC1* were all significantly more methylated in YSTs than in germinomas (Jeyapalan et al. 2011). Furthermore, methylation of the tumor-suppressor gene *RUNX3* in pediatric, but not adult, YSTs has also been identified, which may contribute to the pathogenesis of this childhood GCT subtype (Kato et al. 2003; Furukawa et al. 2009).

Reduced methylation in YST cells in vitro was demonstrated by a study employing the demethylating agent 5-aza-2′-deoxycytidine, a cytidine analog, resulting in reexpression of previously silenced genes such as *SFRP2* (Schönberger et al. 2010). Further investigation of the role of such tumor-suppressor genes, which are methylated in YSTs, may therefore elucidate the cellular mechanisms involved in treatment resistance and identify targets for the development of new therapies.

1.4 Genomic Aberrations

The large majority of studies investigating genomic changes in malignant GCTs have been performed on cases of adult testicular disease, which consistently demonstrate chromosome 12p gain, usually due to isochromosome 12p formation (Atkin and Baker 1982, 1983). This genomic change occurs regardless of histologic subtype. These, and other reports studying expression of genes on 12p, suggest that 12p gain is functionally relevant in TGCTs, by leading to the activation of key stem cell genes, which promote cellular proliferation (Korkola et al. 2006). Pediatric tumors by comparison have been far less intensively studied and generally report small sample numbers. A first study of 16 childhood (predominantly testicular) YSTs showed that the most common changes were gains of chromosomes 1q, 11q, 20q and 22 and loss of 1p, 6q and 16q (Perlman et al. 2000). However, this study identified that 12p gain, characteristic of TGCTs, was only present in one testicular tumor (6 % of the total cases examined) (Perlman et al. 2000). A larger study of 51 cases, including 33 malignant GCTs and 18 teratomas, confirmed similar findings to above, with 12p gain also appearing to be uncommon in childhood forms of the disease (11 % of malignant cases) (Schneider et al. 2001b). A study of childhood and adolescent mediastinal tumors revealed differences by patient age (Schneider et al. 2002). For those patients less than 8 years of age, profiles of malignant mediastinal GCTs were as for testicular and sacrococcygeal tumors and furthermore, 12p gain and abnormalities of the sex chromosomes were not observed (Schneider et al. 2002). By comparison, 12p gain was the most frequent aberration in malignant mediastinal tumors of older children (>8 years of age), with gain of the X chromosome occurring in about half of cases, most likely due to the presence of Klinefelter syndrome, which is known to be associated with an increased risk of developing GCTs (Murray and Nicholson 2010). A study of malignant central nervous system (CNS) GCTs demonstrated that children >8 years and adolescents commonly display 12p gain (Schneider et al. 2006), with genomic aberrations in these tumors being virtually indistinguishable from those in their gonadal or other extragonadal counterparts. This observation strongly favors common pathogenetic mechanisms in the development of both gonadal and extragonadal GCTs (Schneider et al. 2006).

More recently a study of 34 pediatric malignant GCTs, including a relatively large cohort of ovarian tumors ($n = 17$), identified that 12p gain increased in frequency with advancing age, occurring in 29 % of children <5 years of age and in 53 % of those aged between 5 and 16 years (Palmer et al. 2007). However, the significance of 12p gain in malignant GCTs of childhood remains to be elucidated.

Teratomas are generally considered a benign tumor entity; however, in adult oncology practice testicular teratomas often display a greater propensity for growth and malignant transformation than their pediatric counterparts. Tumor biology may account for these observed differences. For example, prepubertal teratomas display normal hybridization profiles, i.e., do not show any or only a few genomic aberrations (Harms et al. 2006; Okpanyi et al. 2011). However, testicular teratomas in postpubertal patients may display isochromosome 12p or more limited regions of 12p gain, as is seen consistently for malignant TGCTs (Schneider et al. 2001b; Harms et al. 2006). Moreover, biologic studies of somatic malignant components that have arisen within malignant teratomas have shown that they retain 12p gain, suggesting that they arise from within the GCT itself (Harms et al. 2006). In contrast, pure ovarian teratomas do not display 12p gain (Kraggerud et al. 2000; Poulos et al. 2006), although "teratomatous" components within malignant ovarian GCTs often do so (Poulos et al. 2006), suggesting two alternative pathways may exist leading to a similar phenotypic appearance (Poulos et al. 2006).

In the last few years, the technique of metaphase-based comparative genomic hybridization (CGH), used to perform the above studies, has been replaced with high-resolution array-based CGH. The latter approach offers the potential to interrogate genomic abnormalities and regions of interest in far greater resolution than has previously been possible. A small study in adult malignant testicular seminomas has been published using this technique

(LeBron et al. 2011), and similar studies in childhood tumors are likely to reveal many novel aberrations that may explain differences in clinical behavior and outcomes, which after validation may in the future be incorporated into prospective clinical trials.

1.5 Transcriptome Analysis

1.5.1 Messenger RNA Profiles Identify Differences Between Pediatric and Adult GCTs

Transcriptomic studies have predominantly been performed in adult TGCTs. A recent review article of 23 such studies highlighted common gene changes in TGCTs, elucidating both transcriptional changes associated with malignant transformation and differentiation patterns of malignant GCTs (Alagaratnam et al. 2011). This study implicated both known (*KRAS, MYCN* and *TPD52*) and novel (*CCT6A, IGFBP3* and *SALL2*) cancer genes in TGCT pathogenesis. Gene expression patterns in malignant GCTs characteristic of ESCs were confirmed (see also Sect. 1.2). A distinctive transcriptomic profile was identified for individual histologic subtypes, particularly striking given that the TGCT genome is largely similar across subtypes (Alagaratnam et al. 2011). Similarly, global mRNA gene expression profiles in pediatric malignant GCTs completely segregate the two main histologic subtypes, YSTs and germinomas, with no differences being observed by anatomic site of disease (Palmer et al. 2008). As seen in adult malignant GCTs, pediatric germinomas were enriched for genes associated with pluripotency, whereas pediatric YSTs were associated with differentiation and proliferation pathways (Palmer et al. 2008) (see Sect. 1.6). Despite this observation, global mRNA profiles segregated pediatric cases from adult malignant TGCTs of the same tumor subtype (Fig. 1.3), suggesting that for mRNA profiles, histologic subtype is the main discriminator and then patient age (Palmer et al. 2008). Thus, the observed differences in the mRNA expression

profiles between pediatric and adult GCTs may presumably be driven at least in part by the alterations in hormonal status that accompany puberty.

1.5.2 MicroRNA Profiles Identify Commonalities Shared by Pediatric and Adult GCTs

Despite their extensive clinical and pathologic heterogeneity, all malignant GCTs are thought to originate from primordial germ cells, but despite this, very few common biologic abnormalities have been identified in these tumors to date. As microRNA profiles have been shown to reflect the development lineage of tumors (Lu et al. 2005), investigating such profiles in malignant GCTs may identify universal biologic findings. MicroRNAs are short, nonprotein-coding RNAs, typically 21–23 nucleotides in length. They post-transcriptionally regulate gene expression through binding sites for the microRNA "seed" region (predominantly nucleotides at positions 2–7 of the microRNA, 2–7nt) in the 3′ untranslated region (3′ UTR) of mRNA targets (Palmer et al. 2010). The first report in GCTs was a genetic screening study for microRNAs that were involved in cellular transformation (Voorhoeve et al. 2006). This identified that miR-372 and miR-373 each allowed proliferation and tumorigenesis in primary human cells containing *RAS* and active wild-type *TP53*, likely through inhibition of the tumor-suppressor gene *LATS2*, preventing *TP53*-mediated *CDK* inhibition. As miR-372 and miR-373 were shown to be overexpressed in TGCTs, these results suggested that these microRNAs contribute to germ cell tumorigenesis by allowing growth even in the presence of wild-type *TP53* (Voorhoeve et al. 2006). A further study used a qRT-PCR platform to interrogate microRNA profiles in adult gonadal GCTs and confirmed overexpression of this oncogenic miR-371~373 cluster (Gillis et al. 2007).

A subsequent microarray study of pediatric malignant GCTs (Palmer et al. 2010) included a reanalysis of the qRT-PCR data described above (Gillis et al. 2007) and extended previous observations. This study identified that all eight

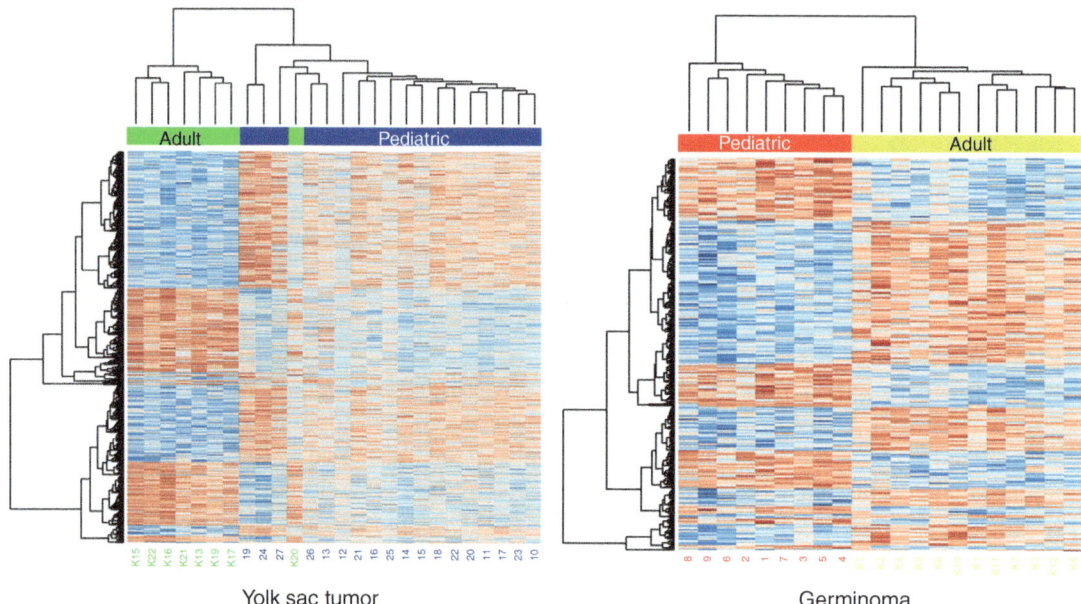

Yolk sac tumor

Germinoma

Fig. 1.3 Comparison of protein-coding mRNA gene expression in pediatric malignant GCTs vs. adult malignant TGCTs. Heatmap comparison of the transcriptional profile of 27 pediatric malignant GCTs with 20 pure (i.e., not mixed) histologic diagnosis adult malignant TGCTs (From Korkola et al. (2006)). Cases from Korkola et al. are given the prefix *K*. The heatmaps are generated from genes identified as differentially expressed within histologic subtypes between the two age groups. Segregation of adult from pediatric yolk sac tumors is shown on the left ($n = 1,180$ genes) and adult from pediatric seminomas on the right ($n = 380$ genes). *Blue*: lower expression. *Red*: higher expression (Reprinted with permission from Palmer et al. (2008))

microRNAs from the miR-371~373 and miR-302 clusters were universally overexpressed in malignant GCTs, regardless of patient age (pediatric or adult), histologic subtype (YST, germinoma, or embryonal carcinoma), or anatomic site of disease (gonadal or extragonadal) (Palmer et al. 2010), consistent with the common origin theory. Indeed, just these eight microRNAs were able to robustly segregate malignant GCTs from nonmalignant samples (benign teratomas and normal gonadal controls) (Fig. 1.4) (Palmer et al. 2010). Furthermore, six of these eight microRNAs from the miR-371~373 and miR-302 clusters shared an identical seed region, suggesting common mRNA targets. Interrogation of samples with matched mRNA expression data confirmed that these microRNAs were biologically significant by globally downregulating mRNA targets which were involved in cancer-associated processes (Palmer et al. 2010). The demonstration that the miR-371~373 cluster was further upregulated in GCT cell lines that had acquired cisplatin resistance (Port et al. 2011) provides further evidence for the functional significance of this microRNA cluster in malignant GCTs. Furthermore, the universal overexpression of the miR-371~373 and miR-302 clusters in malignant GCTs offers translational potential, as they are not reported to be coordinately upregulated in other tumor types or disease states (Palmer et al. 2010). Interestingly, the miR-302 cluster showed further overexpression in YSTs when compared with germinomas, in comparisons of both pediatric and adult GCTs (Murray et al. 2010), and resulted in significant downregulation of mRNA targets, including mediators of key cancer-associated processes, such as tumor-suppressor genes, apoptosis regulators, and transcription factors (Murray et al. 2010). Further evidence of the biologic significance of the miR-302 cluster in malignant GCTs has been provided by a study which confirmed differential regulation of a set of microRNAs between YSTs and germinomas, including the miR-302 cluster, which were predicted to target

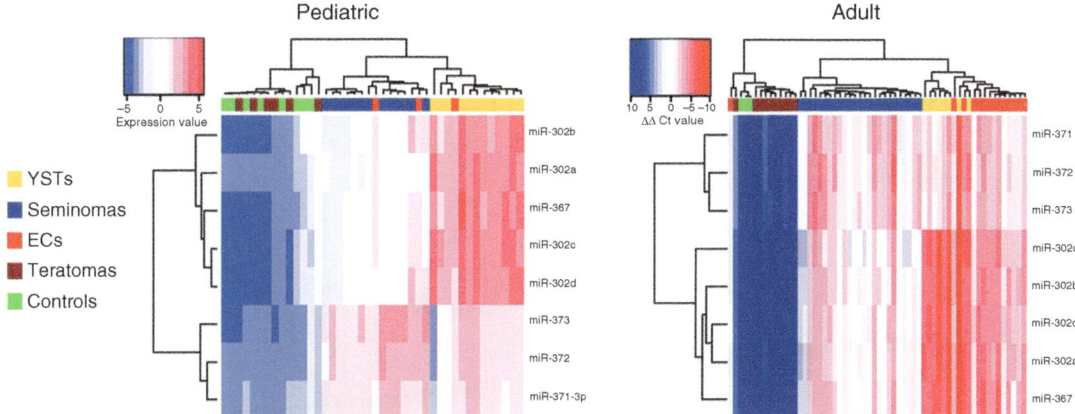

Fig. 1.4 Differential expression of the miR-371~373 and miR-302 clusters in malignant GCTs. Hierarchical clustering analysis based on the miR-371~373 and miR-302 clusters in pediatric (*left*) and adult (*right*) samples. *Blue*: lower expression. *Red*: higher expression (Modified with permission from Palmer and Murray et al. (2010))

the TGF-ß/BMP pathway at multiple sites resulting in bone morphogenetic protein (BMP) signaling pathway activation in YSTs (Fustino et al. 2011). The differential microRNA expression observed is likely to contribute to the relatively aggressive clinical behavior of YSTs and may enable future improvements in clinical diagnosis and treatment.

Regarding clinical diagnosis, the identification of new universal and predictive biomarkers is of significant clinical interest, since the conventional serum tumor markers currently used for detecting and monitoring GCTs, AFP and HCG show limited sensitivity and specificity (Murray and Nicholson 2011) (Göbel et al. 2001). One study identified that levels of all eight members of the miR-371~373 and miR-302 clusters were elevated in the serum at the time of pediatric YST diagnosis, with levels of miR-372 returning to normal during uneventful clinical follow-up (Murray et al. 2011). Subsequently, it was demonstrated that serum levels of microRNAs from these clusters were increased at malignant GCT diagnosis, regardless of patient age, histologic subtype, or anatomic site of disease (Fig. 1.5), with the suggestion of an association with tumor volume at diagnosis (Murray and Coleman 2012). These findings were recently replicated by a group studying stage I adult TGCTs, with increased serum levels of miR-371~373 members demonstrated at the time of malignant GCT diag-

nosis, with levels falling rapidly to normal following definitive treatment with surgical orchidectomy (Belge et al. 2012). Importantly, nine of the 11 cases had "marker-negative" disease, i.e., levels of AFP, HCG, and/or LDH were not raised at diagnosis (Belge et al. 2012). Further larger studies are required to confirm the utility of these microRNAs as potential universal candidate biomarkers in malignant GCTs, which may reduce the need for surveillance imaging for disease monitoring and detection of tumor recurrence in follow-up. In turn, this would decrease the radiation burden and potential risk of second malignancy experienced by patients (Tarin et al. 2009; Silva et al. 2012). Additionally, targeting of microRNAs dysregulated in malignant GCTs may represent a new therapeutic approach, for example, through targeting of *LIN28* (Murray et al. 2013) (see Sect. 1.2).

1.6   Pathway and Proteomic Analysis

During the last decade key insights into the molecular basis of cancer have been elucidated, resulting in a growing understanding of cancer-associated signaling pathways that underlie tumor formation and progression. Among these pathways, *Wnt*, *TGF-beta/BMP*, *PI3K/AKT/mTOR*, *RAS/RAF* and *VEGF* signaling are of special interest, as they

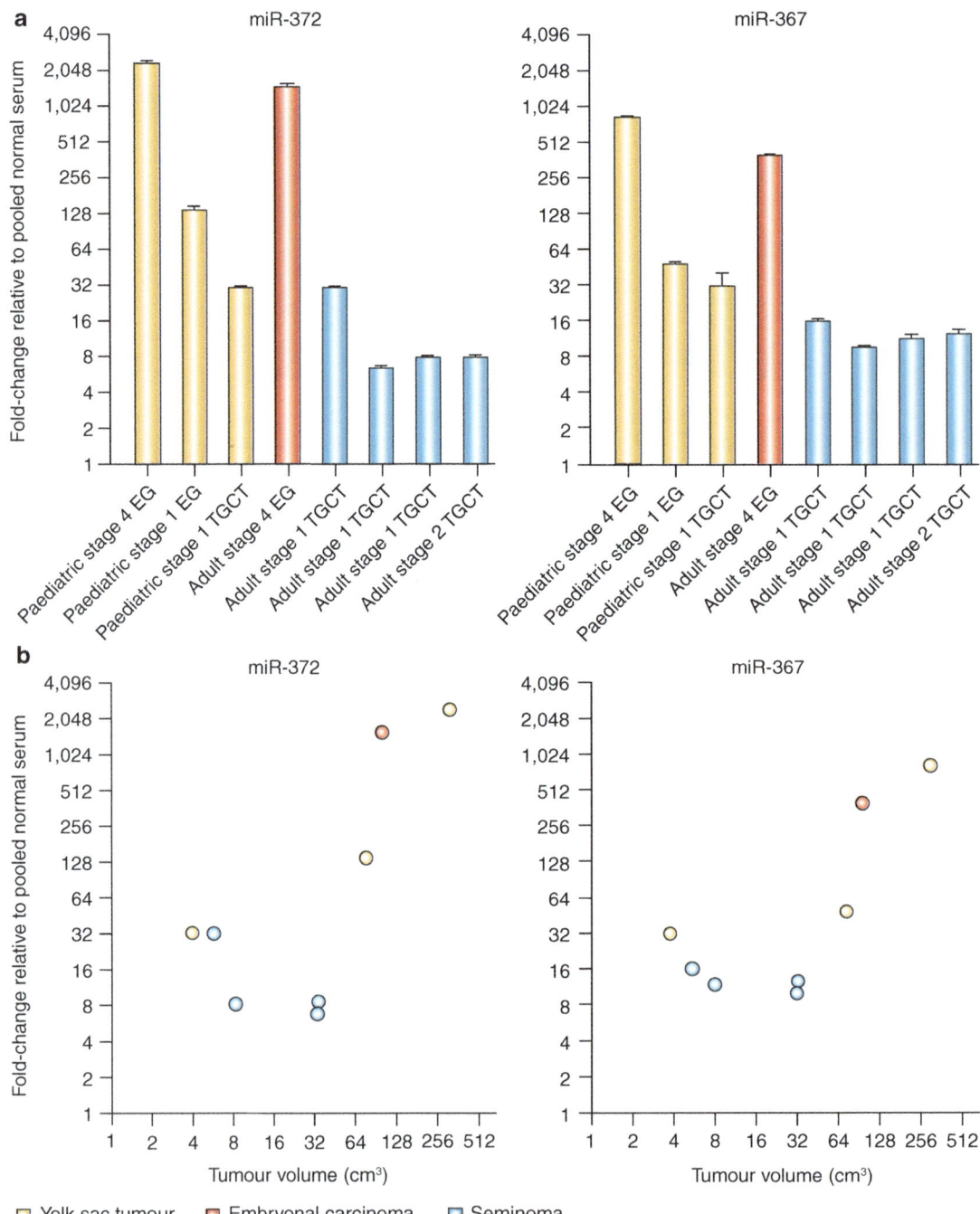

Fig. 1.5 Detection of malignant GCT-associated microRNAs in patient serum. (**a**) Levels of miR-372 from the miR-371~373 cluster (*left*) and miR-367 from the miR-302 cluster (*right*) in the serum at the time of diagnosis in eight MGCTs of different patient age, anatomic site, and histologic subtype. Expression levels are referenced to a pool of control serum samples from six healthy subjects. (**b**) Relationship of serum levels of miR-372 (*left*) and miR-367 (*right*) to total tumor volume at the time of diagnosis for the eight MGCTs described above. *Abbreviations*: *EG* extragonadal, *MGCT* malignant germ cell tumor, *TGCT* testicular germ cell tumor (Reprinted with permission from Murray and Coleman (2012))

have driven the development of a new generation of anticancer drugs, which target specific molecular events/mutations. In malignant GCTs, pathway analysis suggests that some of the cellular mechanisms involved in the pathogenesis of different GCT subtypes are similar. The generally more aggressive clinical behavior of adult in comparison to pediatric malignant GCTs is likely to be due to the differential expression of individual protein-coding genes within those pathways (Palmer et al. 2008). This may potentially be due to different genomic copy number alterations (Palmer et al. 2007) (see Sect. 1.4) or epigenetic mechanisms (see Sect. 1.3) observed between the two age groups of patients.

Expression of genes involved in cancer-associated signaling pathways has been shown to differ between YSTs and dysgerminomas in childhood GCTs. An mRNA microarray analysis identified significant differential expression of genes involved in the *Wnt* pathway between these two subtypes (Fritsch et al. 2006). Besides *WNT13*, *β-catenin* was also differentially expressed and showed nuclear translocation in up to 54 % of YSTs, in contrast to dysgerminomas, where this occurred rarely, indicating activation of *Wnt* signaling in YSTs. Subsequent gene expression analysis of intra- and extracellular regulators of the *Wnt* pathway confirmed the differential expression of several genes between the two subtypes, i.e., *SFRP*s and *DKK1*, predominantly due to epigenetic mechanisms (Schönberger et al. 2010) (see also Sect. 1.3). Methylation of these genes is therefore likely to account for the overexpression of *Wnt/β-catenin* pathway genes seen in YSTs (Fritsch et al. 2006; Palmer et al. 2008). Similarly, differential protein-coding gene expression leads to activation of the *TGF-beta/BMP* pathway in YSTs, in contrast to undifferentiated tumors such as dysgerminomas/seminomas, where *BMP* pathway activity is absent (Fustino et al. 2011). Interestingly, in this study differential microRNA expression between YSTs and dysgerminomas were implicated to account for this observation (see also Sect. 1.5), rather than epigenetic mechanisms.

KIT and its ligand *KITLG* (steel factor) are not only involved in oogenesis and spermatogenesis (see Sect. 1.2) but also known to activate the *PI3K/AKT/mTOR* and *RAS/RAF* pathway, contributing to GCT development. Consequently, recent research has focused on analysis of *KIT/KITLG* in GCTs. Several immunohistochemical studies of adult GCTs detected KIT in seminomas in contrast to non-seminomatous tumors such as YSTs (Bokemeyer et al. 1996; Kemmer et al. 2004; Nakai et al. 2005; Biermann et al. 2007; Nikolaou et al. 2007). Additionally, mutations in *KIT* in codon 816 are associated with the development of bilateral GCTs (Looijenga et al. 2003) and advanced stages of ovarian dysgerminomas (Cheng et al. 2011). Genetic and protein analysis identified different gain-of-function mutations in the *KIT* gene (D816V, D816H) in seminomas resulting in phosphorylation of KIT and PI3K and therefore constitutive activation of the *PI3K* pathway in seminomas, even in the absence of KITLG (Nakai et al. 2005). Unfortunately, although seminomas harboring these activating *KIT* mutations would be predicted to respond to the KIT tyrosine kinase inhibitor imatinib mesylate, results in clinical studies have been disappointing, with no complete or even partial remissions (Einhorn et al. 2006). However, in vitro studies suggest that alternative tyrosine kinase inhibitors, such as dasatinib, may be more promising as treatment options in vivo (Schittenhelm et al. 2006).

In many human cancers, activating mutations in the *KRAS* or *BRAF* gene lead to activation of the *RAS/RAF* pathway resulting in overexpression of its target *MAPK1* (*ERK*). Although *MAPK1* is globally expressed in adult GCTs, activating mutations of *KRAS* and *BRAF* are rare events pointing to the involvement of KIT as an upstream protein in activation of RAS/RAF signaling (McIntyre et al. 2005; Sommerer et al. 2005). Nevertheless, the *BRAF* mutation V600E is detectable in a subgroup of chemotherapy-resistant adult GCTs and significantly correlates with microsatellite instability (Honecker et al. 2009). Of note, this finding could not be verified in a cohort of 70 pediatric GCTs, providing further evidence for age-dependent biologic differences and chromosomal/genetic alterations between pediatric and adult GCTs (Masque-Soler et al. 2012). Therefore, more in vitro and in vivo studies perturbing the specific genes and pathways known to be dysregulated in pediatric malignant

GCTs are needed in order to increase our understanding of this form of the disease.

The cancer signaling pathways described above are involved in protein translation, apoptosis, cell growth, and metabolism. Additionally, *VEGF* signaling results in angiogenesis and in the context of tumor development ensures sufficient nutrient supply to the rapidly dividing tissues. In GCTs, VEGF expression is correlated with metastasis in both seminomatous and non-seminomatous adult TGCTs (Fukuda et al. 1999), assigning VEGF a central role in the pathogenesis of advanced and invasive tumor stages. Interestingly, VEGF expression is significantly higher in teratomas compared to other TGCTs or healthy tissue (Jones et al. 2000). As a result, the VEGF antagonist bevacizumab was successfully employed in combination with other chemotherapy agents in a case of growing teratoma syndrome in an adolescent, allowing subsequent surgical resection (Calaminus et al. 2009).

Although there is increasing insight into the molecular mechanisms of cancer signaling pathways, it is important to view them as an integrated network in the initial development and subsequent growth of cancer. Furthermore, factors such as the tumor microenvironment have recently been shown to be important and are likely to alter the protein expression within tumor cells. Thus, the identification of key molecular targets and the establishment of novel therapeutic agents in pediatric as well as adult malignant GCTs require further functional analysis in vitro and in vivo. Future studies integrating mRNA and microRNA expression profiles in malignant GCTs from patients with good and adverse clinical outcomes are likely to identify regulatory pathways and networks associated with treatment sensitivity and resistance.

References

Addeo R, Crisci S, D'Angelo V, Vincenzi B, Casale F, Pettinato G et al (2007) Bax mutation and overexpression inversely correlate with immature phenotype and prognosis of childhood germ cell tumors. Oncol Rep 17:1155–1161

Alagaratnam S, Lind GE, Kraggerud SM, Lothe RA, Skotheim RI (2011) The testicular germ cell tumour transcriptome. Int J Androl 34:e133–e150; discussion e50–e51

Atkin NB, Baker MC (1982) Specific chromosome change, i(12p), in testicular tumours? Lancet 2:1349

Atkin NB, Baker MC (1983) i(12p): specific chromosomal marker in seminoma and malignant teratoma of the testis? Cancer Genet Cytogenet 10:199–204

Bamberg M, Kortmann RD, Calaminus G, Becker G, Meisner C, Harms D et al (1999) Radiation therapy for intracranial germinoma: results of the German cooperative prospective trials MAKEI 83/86/89. J Clin Oncol 17:2585–2592

Belge G, Dieckmann KP, Spiekermann M, Balks T, Bullerdiek J (2012) Serum levels of microRNAs miR-371-3: a novel class of serum biomarkers for testicular germ cell tumors? Eur Urol 61:1068–1069

Biermann K, Heukamp LC, Steger K, Zhou H, Franke FE, Sonnack V et al (2007) Genome-wide expression profiling reveals new insights into pathogenesis and progression of testicular germ cell tumors. Cancer Genomics Proteomics 4:359–367

Bokemeyer C, Kuczyk MA, Dunn T, Serth J, Hartmann K, Jonasson J et al (1996) Expression of stem-cell factor and its receptor c-kit protein in normal testicular tissue and malignant germ-cell tumours. J Cancer Res Clin Oncol 122:301–306

Bussey KJ, Lawce HJ, Himoe E, Shu XO, Heerema NA, Perlman EJ et al (2001) SNRPN methylation patterns in germ cell tumors as a reflection of primordial germ cell development. Genes Chromosomes Cancer 32:342–352

Calaminus G, Bamberg M, Harms D, Jürgens H, Kortmann RD, Sorensen N et al (2005) AFP/beta-HCG secreting CNS germ cell tumors: long-term outcome with respect to initial symptoms and primary tumor resection. Results of the cooperative trial MAKEI 89. Neuropediatrics 36:71–77

Calaminus G, Schneider DT, Weissbach L, Schönberger S, Okpanyi V, Leuschner I et al (2009) Survival after an antiangiogenetic therapy and surgery in a wide spread growing teratoma originating from a testicular mixed malignant germ cell tumor. Klin Padiatr 221:136–140

Cao D, Allan RW, Cheng L, Peng Y, Guo CC, Dahiya N et al (2011a) RNA-binding protein LIN28 is a marker for testicular germ cell tumors. Hum Pathol 42:710–718

Cao D, Liu A, Wang F, Allan RW, Mei K, Peng Y et al (2011b) RNA-binding protein LIN28 is a marker for primary extragonadal germ cell tumors: an immunohistochemical study of 131 cases. Mod Pathol 24:288–296

Cheng L, Roth LM, Zhang S, Wang M, Morton MJ, Zheng W et al (2011) KIT gene mutation and amplification in dysgerminoma of the ovary. Cancer 117:2096–2103

Doitsidou M, Reichman-Fried M, Stebler J, Koprunner M, Dorries J, Meyer D et al (2002) Guidance of primordial germ cell migration by the chemokine SDF-1. Cell 111:647–659

Einhorn LH, Brames MJ, Heinrich MC, Corless CL, Madani A (2006) Phase II study of imatinib mesylate in chemotherapy refractory germ cell tumors expressing KIT. Am J Clin Oncol 29:12–13

Fritsch MK, Schneider DT, Schuster AE, Murdoch FE, Perlman EJ (2006) Activation of Wnt/beta-catenin signaling in distinct histologic subtypes of human germ cell tumors. Pediatr Dev Pathol 9:115–131

Fukuda S, Shirahama T, Imazono Y, Tsushima T, Ohmori H, Kayajima T et al (1999) Expression of vascular endothelial growth factor in patients with testicular germ cell tumors as an indicator of metastatic disease. Cancer 85:1323–1330

Furukawa S, Haruta M, Arai Y, Honda S, Ohshima J, Sugawara W et al (2009) Yolk sac tumor but not seminoma or teratoma is associated with abnormal epigenetic reprogramming pathway and shows frequent hypermethylation of various tumor suppressor genes. Cancer Sci 100:698–708

Fustino N, Rakheja D, Ateek CS, Neumann JC, Amatruda JF (2011) Bone morphogenetic protein signalling activity distinguishes histological subsets of paediatric germ cell tumours. Int J Androl 34:e218–e233

Gilbert DC, Chandler I, McIntyre A, Goddard NC, Gabe R, Huddart RA et al (2009) Clinical and biological significance of CXCL12 and CXCR4 expression in adult testes and germ cell tumours of adults and adolescents. J Pathol 217:94–102

Gillis AJ, Stoop HJ, Hersmus R, Oosterhuis JW, Sun Y, Chen C et al (2007) High-throughput microRNAome analysis in human germ cell tumours. J Pathol 213:319–328

Gillis AJ, Stoop H, Biermann K, van Gurp RJ, Swartzman E, Cribbes S et al (2011) Expression and interdependencies of pluripotency factors LIN28, OCT3/4, NANOG and SOX2 in human testicular germ cells and tumours of the testis. Int J Androl 34:e160–e174

Göbel U, Schneider DT, Calaminus G, Jürgens H, Spaar HJ, Sternschulte W et al (2001) Multimodal treatment of malignant sacrococcygeal germ cell tumors: a prospective analysis of 66 patients of the German cooperative protocols MAKEI 83/86 and 89. J Clin Oncol 19:1943–1950

Göbel U, Calaminus G, Schneider DT, Schmidt P, Haas RJ (2002) Management of germ cell tumors in children: approaches to cure. Onkologie 25:14–22

Gu Y, Runyan C, Shoemaker A, Surani A, Wylie C (2009) Steel factor controls primordial germ cell survival and motility from the time of their specification in the allantois, and provides a continuous niche throughout their migration. Development 136:1295–1303

Gu Y, Runyan C, Shoemaker A, Surani MA, Wylie C (2011) Membrane-bound steel factor maintains a high local concentration for mouse primordial germ cell motility, and defines the region of their migration. PLoS One 6:e25984

Harms D, Zahn S, Göbel U, Schneider DT (2006) Pathology and molecular biology of teratomas in childhood and adolescence. Klin Padiatr 218:296–302

Hoffman HJ, Otsubo H, Hendrick EB, Humphreys RP, Drake JM, Becker LE et al (1991) Intracranial germ-cell tumors in children. J Neurosurg 74:545–551

Honecker F, Wermann H, Mayer F, Gillis AJ, Stoop H, van Gurp RJ et al (2009) Microsatellite instability, mismatch repair deficiency, and BRAF mutation in treatment-resistant germ cell tumors. J Clin Oncol 27:2129–2136

Honorio S, Agathanggelou A, Wernert N, Rothe M, Maher ER, Latif F (2003) Frequent epigenetic inactivation of the RASSF1A tumour suppressor gene in testicular tumours and distinct methylation profiles of seminoma and nonseminoma testicular germ cell tumours. Oncogene 22:461–466

Jeyapalan JN, Noor DA, Lee SH, Tan CL, Appleby VA, Kilday JP et al (2011) Methylator phenotype of malignant germ cell tumours in children identifies strong candidates for chemotherapy resistance. Br J Cancer 105:575–585

Jones A, Fujiyama C, Turner K, Fuggle S, Cranston D, Turley H et al (2000) Angiogenesis and lymphangiogenesis in stage 1 germ cell tumours of the testis. BJU Int 86:80–86

Kaatsch P (2004) German Childhood Cancer Registry and its favorable setting. Bundesgesundheitsblatt Gesundheitsforschung Gesundheitsschutz 47:437–443

Kato N, Tamura G, Fukase M, Shibuya H, Motoyama T (2003) Hypermethylation of the RUNX3 gene promoter in testicular yolk sac tumor of infants. Am J Pathol 163:387–391

Kato N, Shibuya H, Fukase M, Tamura G, Motoyama T (2006) Involvement of adenomatous polyposis coli (APC) gene in testicular yolk sac tumor of infants. Hum Pathol 37:48–53

Kemmer K, Corless CL, Fletcher JA, McGreevey L, Haley A, Griffith D et al (2004) KIT mutations are common in testicular seminomas. Am J Pathol 164:305–313

Korkola JE, Houldsworth J, Chadalavada RS, Olshen AB, Dobrzynski D, Reuter VE et al (2006) Downregulation of stem cell genes, including those in a 200-kb gene cluster at 12p13.31, is associated with in vivo differentiation of human male germ cell tumors. Cancer Res 66:820–827

Koul S, McKiernan JM, Narayan G, Houldsworth J, Bacik J, Dobrzynski DL et al (2004) Role of promoter hypermethylation in Cisplatin treatment response of male germ cell tumors. Mol Cancer 3:16

Kraggerud SM, Szymanska J, Abeler VM, Kaern J, Eknaes M, Heim S et al (2000) DNA copy number changes in malignant ovarian germ cell tumors. Cancer Res 60:3025–3030

LeBron C, Pal P, Brait M, Dasgupta S, Guerrero-Preston R, Looijenga LH et al (2011) Genome-wide analysis of genetic alterations in testicular primary seminoma using high resolution single nucleotide polymorphism arrays. Genomics 97:341–349

Lee SH, Appleby V, Jeyapalan JN, Palmer RD, Nicholson JC, Sottile V et al (2011) Variable methylation of the imprinted gene, SNRPN, supports a relationship between intracranial germ cell tumours and neural stem cells. J Neurooncol 101:419–428

Looijenga LH, de Leeuw H, van Oorschot M, van Gurp RJ, Stoop H, Gillis AJ et al (2003) Stem cell factor receptor (c-KIT) codon 816 mutations predict development of bilateral testicular germ-cell tumors. Cancer Res 63:7674–7678

Lu J, Getz G, Miska EA, Alvarez-Saavedra E, Lamb J, Peck D et al (2005) MicroRNA expression profiles classify human cancers. Nature 435:834–838

Lu TY, Lu RM, Liao MY, Yu J, Chung CH, Kao CF et al (2010) Epithelial cell adhesion molecule regulation is associated with the maintenance of the undifferentiated phenotype of human embryonic stem cells. J Biol Chem 285:8719–8732

Mann JR, Raafat F, Robinson K, Imeson J, Gornall P, Sokal M et al (2000) The United Kingdom Children's Cancer Study Group's second germ cell tumor study: carboplatin, etoposide, and bleomycin are effective treatment for children with malignant extracranial germ cell tumors, with acceptable toxicity. J Clin Oncol 18:3809–3818

Marina N, London WB, Frazier AL, Lauer S, Rescorla F, Cushing B et al (2006) Prognostic factors in children with extragonadal malignant germ cell tumors: a pediatric intergroup study. J Clin Oncol 24: 2544–2548

Masque-Soler N, Szczepanowski M, Leuschner I, Vokuhl C, Haag J, Calaminus G et al (2012) Absence of BRAF mutation in pediatric and adolescent germ cell tumors indicate biological differences to adult tumors. Pediatr Blood Cancer 59:732–735

McIntyre A, Summersgill B, Spendlove HE, Huddart R, Houlston R, Shipley J (2005) Activating mutations and/or expression levels of tyrosine kinase receptors GRB7, RAS, and BRAF in testicular germ cell tumors. Neoplasia 7:1047–1052

Molyneaux KA, Zinszner H, Kunwar PS, Schaible K, Stebler J, Sunshine MJ et al (2003) The chemokine SDF1/CXCL12 and its receptor CXCR4 regulate mouse germ cell migration and survival. Development 130:4279–4286

Murray MJ, Coleman N (2012) Testicular cancer: a new generation of biomarkers for malignant germ cell tumours. Nat Rev Urol 9:298–300

Murray MJ, Nicholson JC (2010) Germ cell tumours in children and adolescents. Paediatr Child Health 20: 109–116

Murray MJ, Nicholson JC (2011) Alpha-fetoprotein. Arch Dis Child Educ Pract Ed 96:141–147

Murray MJ, Fern LA, Stark DP, Eden TO, Nicholson JC (2009) Breaking down barriers: improving outcomes for teenagers and young adults with germ cell tumours. Oncol Rev 3:201–206

Murray MJ, Saini HK, van Dongen S, Palmer RD, Muralidhar B, Pett MR et al (2010) The two most common histological subtypes of malignant germ cell tumour are distinguished by global microRNA profiles, associated with differential transcription factor expression. Mol Cancer 9:290

Murray MJ, Halsall DJ, Hook CE, Williams DM, Nicholson JC, Coleman N (2011) Identification of microRNAs from the miR-371~373 and miR-302 clusters as potential serum biomarkers of malignant germ cell tumors. Am J Clin Pathol 135:119–125

Murray MJ, Saini HK, Siegler CA, Hanning JE, Barker EM, van Dongen S et al (2013) LIN28 expression in malignant germ cell tumors downregulates let-7 and increases oncogene levels. Cancer Res 17th June, Epub ahead of print

Nakai Y, Nonomura N, Oka D, Shiba M, Arai Y, Nakayama M et al (2005) KIT (c-kit oncogene product) pathway is constitutively activated in human testicular germ cell tumors. Biochem Biophys Res Commun 337:289–296

Ng VY, Ang SN, Chan JX, Choo AB (2010) Characterization of epithelial cell adhesion molecule as a surface marker on undifferentiated human embryonic stem cells. Stem Cells 28:29–35

Nicholson JC, Palmer RD (2010) Germ cell tumors. In: Estlin EJ, Gilbertson RJ, Wynn RF (eds) Pediatric hematology and oncology. Wiley-Blackwell Publishing Ltd, Oxford, pp 275–305

Nikolaou M, Valavanis C, Aravantinos G, Fountzilas G, Tamvakis N, Lekka I et al (2007) Kit expression in male germ cell tumors. Anticancer Res 27: 1685–1688

Okpanyi V, Schneider DT, Zahn S, Sievers S, Calaminus G, Nicholson JC et al (2011) Analysis of the adenomatous polyposis coli (APC) gene in childhood and adolescent germ cell tumors. Pediatr Blood Cancer 56:384–391

Palmer RD, Foster NA, Vowler SL, Roberts I, Thornton CM, Hale JP et al (2007) Malignant germ cell tumours of childhood: new associations of genomic imbalance. Br J Cancer 96:667–676

Palmer RD, Barbosa-Morais NL, Gooding EL, Muralidhar B, Thornton CM, Pett MR et al (2008) Pediatric malignant germ cell tumors show characteristic transcriptome profiles. Cancer Res 68:4239–4247

Palmer RD, Murray MJ, Saini HK, van Dongen S, Abreu-Goodger C, Muralidhar B et al (2010) Malignant germ cell tumors display common microRNA profiles resulting in global changes in expression of messenger RNA targets. Cancer Res 70:2911–2923

Perlman EJ, Hu J, Ho D, Cushing B, Lauer S, Castleberry RP (2000) Genetic analysis of childhood endodermal sinus tumors by comparative genomic hybridization. J Pediatr Hematol Oncol 22:100–105

Port M, Glaesener S, Ruf C, Riecke A, Bokemeyer C, Meineke V et al (2011) Micro-RNA expression in cis-platin resistant germ cell tumor cell lines. Mol Cancer 10:52

Poulos C, Cheng L, Zhang S, Gersell DJ, Ulbright TM (2006) Analysis of ovarian teratomas for isochromosome 12p: evidence supporting a dual histogenetic pathway for teratomatous elements. Mod Pathol 19:766–771

Poveda A, Kaye SB, McCormack R, Wang S, Parekh T, Ricci D et al (2011) Circulating tumor cells predict progression free survival and overall survival in patients with relapsed/recurrent advanced ovarian cancer. Gynecol Oncol 122:567–572

Schittenhelm MM, Shiraga S, Schroeder A, Corbin AS, Griffith D, Lee FY et al (2006) Dasatinib (BMS-354825), a dual SRC/ABL kinase inhibitor, inhibits the kinase activity of wild-type, juxtamembrane, and activation loop mutant KIT isoforms associated with human malignancies. Cancer Res 66:473–481

Schneider DT, Schuster AE, Fritsch MK, Hu J, Olson T, Lauer S et al (2001a) Multipoint imprinting analysis indicates a common precursor cell for gonadal and nongonadal pediatric germ cell tumors. Cancer Res 61:7268–7276

Schneider DT, Schuster AE, Fritsch MK, Calaminus G, Harms D, Göbel U et al (2001b) Genetic analysis of childhood germ cell tumors with comparative genomic hybridization. Klin Padiatr 213:204–211

Schneider DT, Schuster AE, Fritsch MK, Calaminus G, Göbel U, Harms D et al (2002) Genetic analysis of mediastinal nonseminomatous germ cell tumors in children and adolescents. Genes Chromosomes Cancer 34:115–125

Schneider DT, Calaminus G, Koch S, Teske C, Schmidt P, Haas RJ et al (2004) Epidemiologic analysis of 1,442 children and adolescents registered in the German germ cell tumor protocols. Pediatr Blood Cancer 42:169–175

Schneider DT, Zahn S, Sievers S, Alemazkour K, Reifenberger G, Wiestler OD et al (2006) Molecular genetic analysis of central nervous system germ cell tumors with comparative genomic hybridization. Mod Pathol 19:864–873

Schönberger S, Okpanyi V, Alemazkour K, Looijenga LH, Nicholson JC, Borkhardt A et al (2010) Extracellular regulators of the WNT signalling pathway in childhood germ cell tumors: methylation of the SFRP2 promoter leads to WNT activation and ß-Catenin accumulation. Pediatr Blood Cancer 55:804

Schönberger S, Okpanyi V, Calaminus G, Heikaus S, Leuschner I, Nicholson JC et al (2013) EPCAM-A novel molecular target for the treatment of pediatric and adult germ cell tumors. Genes Chromosomes Cancer 52(1):24–32

Sievers S, Alemazkour K, Zahn S, Perlman EJ, Gillis AJ, Looijenga LH et al (2005) IGF2/H19 imprinting analysis of human germ cell tumors (GCTs) using the methylation-sensitive single-nucleotide primer extension method reflects the origin of GCTs in different stages of primordial germ cell development. Genes Chromosomes Cancer 44:256–264

Silva MV, Motamedinia P, Badalato GM, Hruby G, McKiernan JM (2012) Diagnostic radiation exposure risk in a contemporary cohort of male patients with germ cell tumor. J Urol 187:482–486

Sommerer F, Hengge UR, Markwarth A, Vomschloss S, Stolzenburg JU, Wittekind C et al (2005) Mutations of BRAF and RAS are rare events in germ cell tumours. Int J Cancer 113:329–335

Stang A, Trabert B, Wentzensen N, Cook MB, Rusner C, Oosterhuis JW et al (2012) Gonadal and extragonadal germ cell tumours in the United States, 1973–2007. Int J Androl 35:616–625

Stanley MA, Pett MR, Coleman N (2007) HPV: from infection to cancer. Biochem Soc Trans 35: 1456–1460

Staton AA, Knaut H, Giraldez AJ (2011) miRNA regulation of Sdf1 chemokine signaling provides genetic robustness to germ cell migration. Nat Genet 43: 204–211

Tarin TV, Sonn G, Shinghal R (2009) Estimating the risk of cancer associated with imaging related radiation during surveillance for stage I testicular cancer using computerized tomography. J Urol 181:627–632; discussion 32–33

Teilum G (1965) Classification of endodermal sinus tumour (mesoblatoma vitellinum) and so-called "embryonal carcinoma" of the ovary. Acta Pathol Microbiol Scand 64:407–429

Viswanathan SR, Daley GQ, Gregory RI (2008) Selective blockade of microRNA processing by Lin28. Science 320:97–100

Viswanathan SR, Powers JT, Einhorn W, Hoshida Y, Ng TL, Toffanin S et al (2009) Lin28 promotes transformation and is associated with advanced human malignancies. Nat Genet 41:843–848

Voorhoeve PM, le Sage C, Schrier M, Gillis AJ, Stoop H, Nagel R et al (2006) A genetic screen implicates miRNA-372 and miRNA-373 as oncogenes in testicular germ cell tumors. Cell 124:1169–1181

West JA, Viswanathan SR, Yabuuchi A, Cunniff K, Takeuchi A, Park IH et al (2009) A role for Lin28 in primordial germ-cell development and germ-cell malignancy. Nature 460:909–913

Xue D, Peng Y, Wang F, Allan RW, Cao D (2011) RNA-binding protein LIN28 is a sensitive marker of ovarian primitive germ cell tumours. Histopathology 59: 452–459

Epidemiology of Germ Cell Tumors

2

Jenny N. Poynter

Contents

J.N. Poynter
Department of Pediatrics, University of Minnesota,
Minneapolis, MN, USA
e-mail: poynt006@umn.edu

2.1 Incidence and Survival

Comparing incidence and survival, particularly across geographic locations and demographic characteristics, is often useful for generating hypotheses about risk factors for disease. These descriptive analyses can be used to identify differences in person, place, and/or time that epidemiologists can then investigate in future studies with individual level data.

Two peaks in incidence are typically observed for GCTs in both males and females, with the first occurring between the ages of 0–4 years and the second occurring following puberty (Figs. 2.1 and 2.2). In males, the incidence in adolescence and early adulthood is much higher than in the pediatric age group with a peak in the late 20s and early 30s (Fig. 2.1). In females, the incidence in adolescence is similar to the incidence in the pediatric age group and, malignant GCTs are very rare in adult women (Fig. 2.2). This is thought to be due to the more limited number of germ cells in females in the mature ovaries (Moller and Evans 2003; Motta et al. 1997).

A.L. Frazier, J.F. Amatruda (eds.), *Pediatric Germ Cell Tumors*, Pediatric Oncology 1,
DOI 10.1007/978-3-642-38971-9_2, © Springer-Verlag Berlin Heidelberg 2014

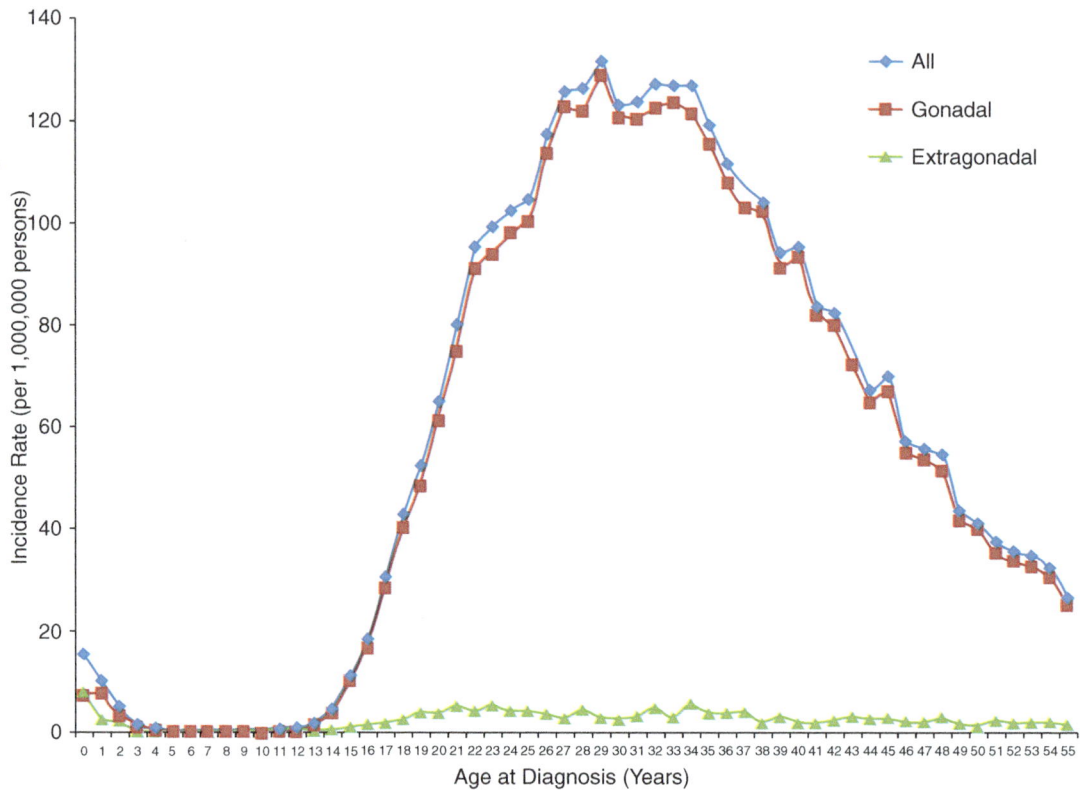

Fig. 2.1 Age-specific incidence of germ cell tumors (per 1,000,000) by tumor location in males in the United States, 1975–2008 (*Source*: Horner et al. 2011)

Incidence patterns differ by tumor location in boys and girls in the pediatric age group, with the tumors occurring with equal frequency in the testes and extragonadal locations in boys, while the tumors in girls are almost exclusively in extragonadal locations prior to the age of 4 years (Figs. 2.1 and 2.2). Several factors may contribute to these differences. The higher rate of gonadal GCTs in young boys may reflect a more permissive environment in the immature testis than in the immature ovary. Another factor may be physiologic differences between the sexes: in females, germ cells undergo a prenatal meiotic arrest that persists until puberty, whereas in males, mitotic proliferationx of germ cells resumes shortly after birth and continues throughout childhood (Wilhelm et al. 2007).

The distribution of tumors by location differs in the pediatric and adult age groups, with extragonadal tumors comprising a larger percentage of tumors diagnosed in children before the age of 4 years than individuals diagnosed after age 10 years. Previous reports have estimated that 40–55 % of pediatric GCTs are found in extragonadal locations (Bernstein et al. 1999; Chen et al. 2005a; De Backer et al. 2008; Harms and Janig 1986; Shu et al. 1995), while only 5–10 % of GCTs in adults are found in extragonadal locations (Houldsworth et al. 2006; Rescorla and Breitfeld 1999). This difference is hypothesized to be due to differences in the maturity of the germ cells that give rise to the tumors in these age groups (Oosterhuis et al. 2007). Pediatric GCTs likely originate from a PGC that underwent immediate reprogramming to become a pluripotent embryonic germ cell, while GCTs in adolescents and young adults most likely originate from more mature PGCs (Oosterhuis and Looijenga 2005), which may be unable to survive outside of the normal niches of the ovary and testis.

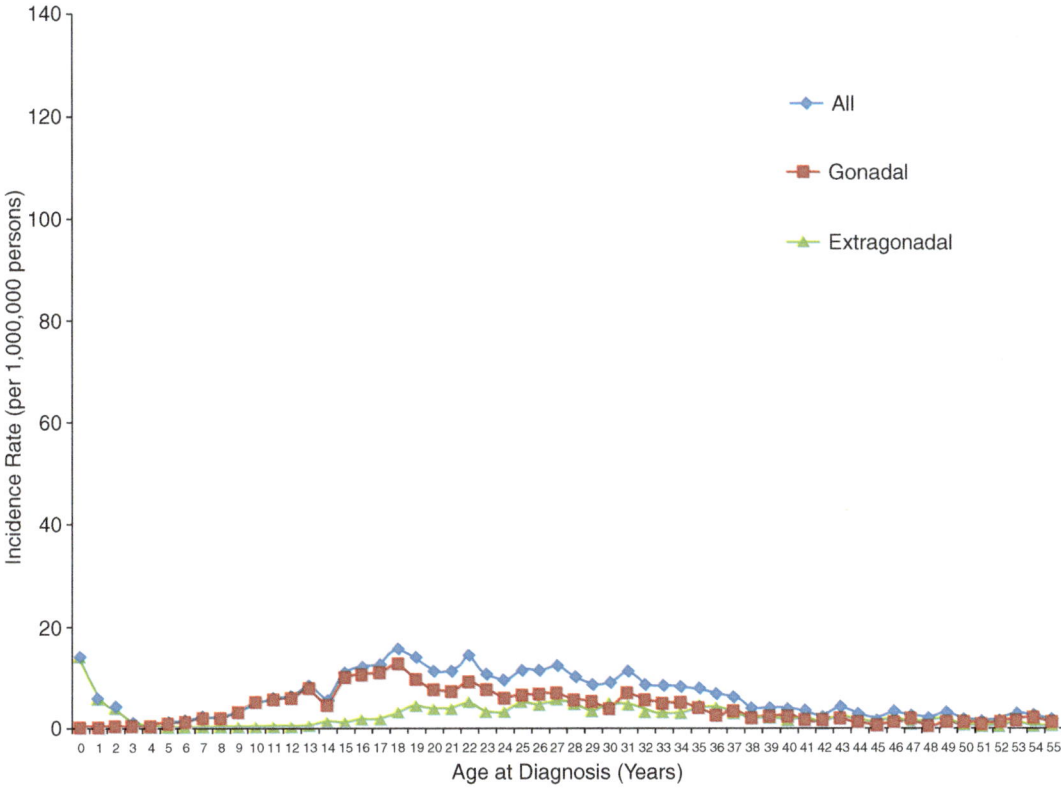

Fig. 2.2 Age-specific incidence of germ cell tumors (per 1,000,000) by tumor location in females in the United States, 1975–2008 (*Source*: Horner et al. 2011)

2.2 International Incidence Rates

2.2.1 Pediatric GCTs

As shown in Figs. 2.3 and 2.4, considerable variation in incidence rates for pediatric GCTs is observed worldwide for both boys and girls (Parkin et al. 1998). These rates include all histologic subtypes of germ cell tumors in all anatomic locations in children ages 0–14 years. Rates were not included for registries where fewer than ten cases were reported. In boys, the highest incidence was observed in Maori of New Zealand (9.7/million); Osaka, Japan (9.2/million); and Chinese (8.8/million) and Malay (7.8/million) in Singapore. The lowest rates in boys were observed in Egypt (1.4/million), Thailand (1.4/million), and Bulgaria (1.6/million) with intermediate rates

in Norway (5.9/million), Sweden (5.3/million), Denmark (5.1/million), and US whites (3.5/million). In girls, seven registries reported incidence greater than 7/million (Osaka, Japan; Seoul, Korea; Chinese of Singapore; US blacks; Japan; Tianjin, China; non-Maori of New Zealand), while the remaining registries reported incidence less than 6/million. With the exception of Egypt, all of the cancer registries in Africa reported fewer than ten cases and were not included in this report. These data must be interpreted with caution because cancer registries differ across countries with respect to scope, completeness, and overall number of cases observed. However, the rates were highest among children with Asian ancestry in both boys and girls, particularly in Japan, Singapore, China, and New Zealand. These observations could suggest a genetic or environmental contribution to GCT risk.

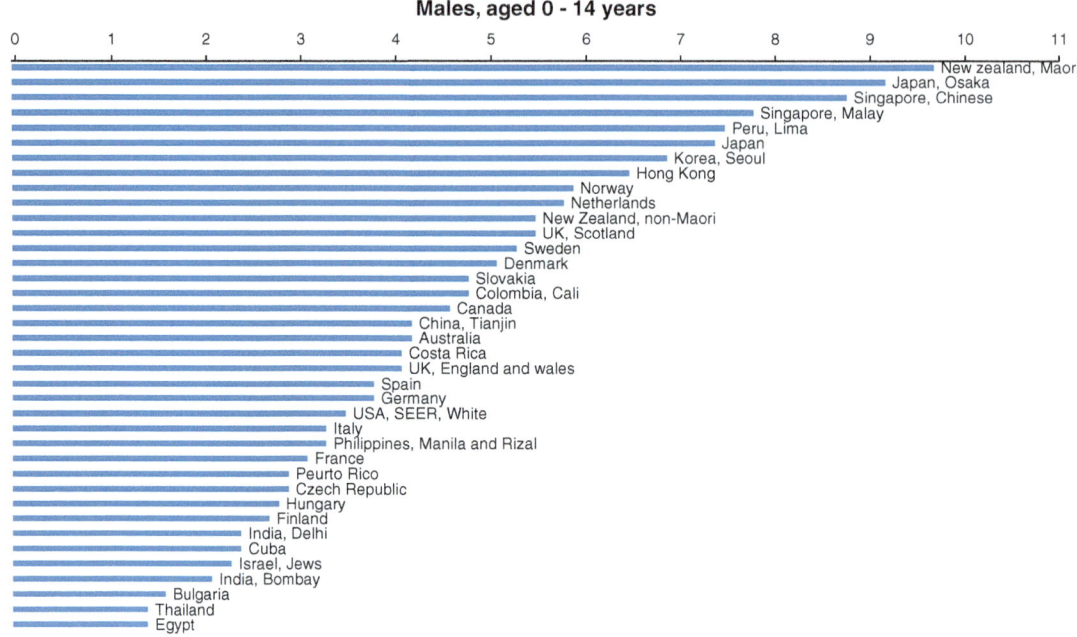

Fig. 2.3 International age-adjusted incidence rates of germ cell tumors (per 1,000,000) in boys aged 0–14 years. All rates are standardized to the world standard population (*Source*: Parkin et al. 1998)

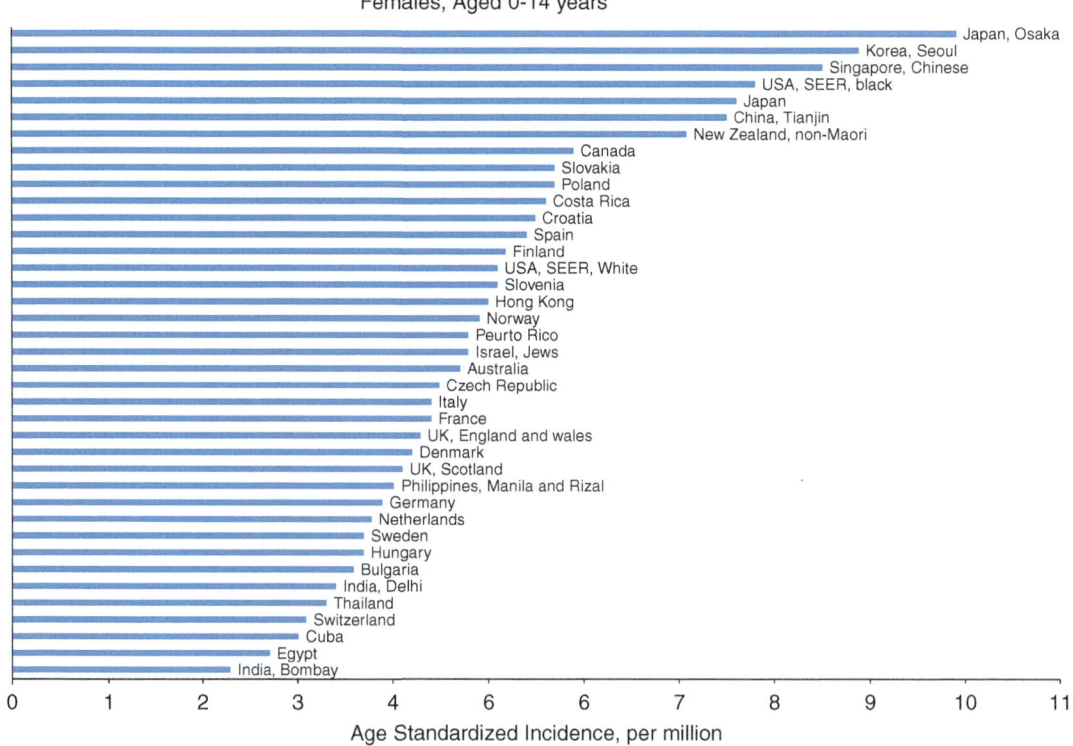

Fig. 2.4 International age-adjusted incidence rates of germ cell tumors (per 1,000,000) in girls aged 0–14 years. All rates are standardized to the world standard population (*Source*: Parkin et al. 1998)

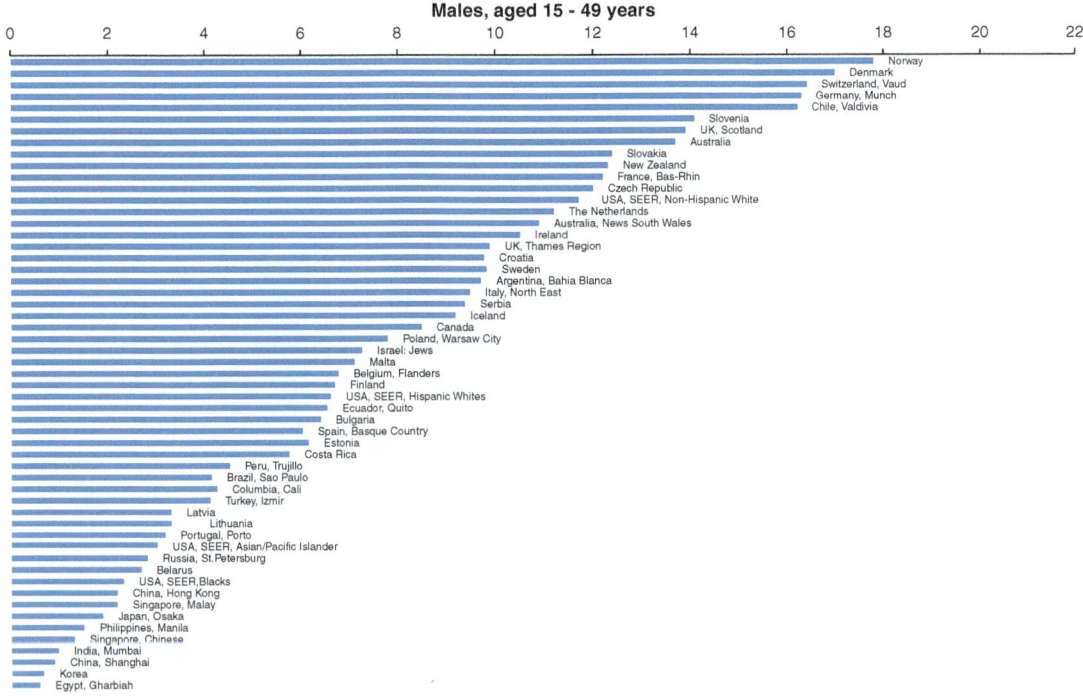

Fig. 2.5 International age-adjusted incidence rates of testicular cancer (per 100,000) in males aged 15–49 years. All rates are standardized to the world standard population (*Source*: Curado et al. 2007)

2.2.2 Adult Testicular GCTs

The incidence of testicular GCT varies widely throughout the world, with rates ranging from 0.6 per 100,000 men in Gharbiyah, Egypt, to 17.8 per 100,000 men in Norway (Fig. 2.5). These rates include testicular cancers in men aged 15–49 years. Overall, rates were highest in northern and western European countries and lowest in African and Asian countries. Central and South American countries tended to have intermediate rates with the exception of Valdivia, Chile, which reported one of the highest incidence rates in the world (16.2 per 100,000 men). The incidence in racial and ethnic subgroups in the United States mirrored international rates, with relatively high rates in non-Hispanic whites (11.7 per 100,000), intermediate rates in Hispanic whites (6.6 per 100,000), and low rates in black and Asian men (2.3 and 3.0 per 100,000 men, respectively). The variation in incidence rates by country has been reported on extensively in Europe

(Bray et al. 2006a, b; Jacobsen et al. 2006) and on a more limited basis worldwide (Chia et al. 2010; Nicolaides et al. 1994; Purdue et al. 2005). The differing incidence has been hypothesized to be due to an as yet unidentified risk factor or factors with differing prevalence across countries (Bray et al. 2006b). Genetic susceptibility may also explain some of the variation.

The low rates of TGCT in Asian men in adulthood are in stark contrast to the incidence rates in the pediatric age group, where several registries in Asia reported the highest incidence rates in the world. Part of the explanation for this difference could be due to the inclusion of intracranial and other extragonadal GCTs in the pediatric incidence rates, especially considering the high rates of intracranial GCT reported in Asia (Kuratsu and Ushio 1996; Parkin et al. 1998; Wong et al. 2005). However, this is unlikely to explain the entire difference as the four highest rates of testicular GCTs in boys were reported in Chinese in Singapore (5.8/million); Lima, Peru (5.6/million); Maori of New Zealand (5.4/million); and

Fig. 2.6 Incidence rates of GCT by tumor histology in boys and girls aged 0–14 years in the United States, 1975–2008. Rates are standardized to the 2000 US standard population (*Source*: Horner et al. 2011)

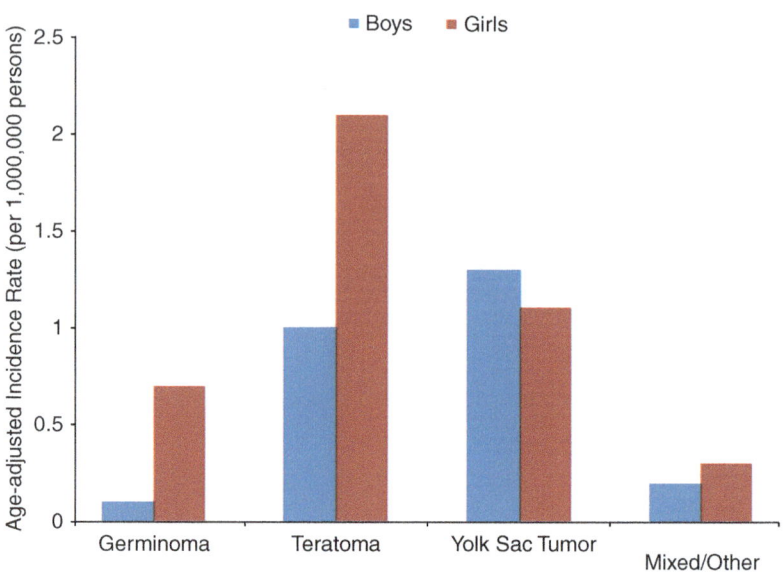

Seoul, Korea (4.6/million) (Parkin et al. 1998). Further studies will be required to evaluate this interesting finding.

2.2.3   Adult Ovarian GCTs

Malignant ovarian GCTs are very rare in adult women, with rates below 1 per 100,000 women in every cancer registry reporting data for IARC's Cancer Incidence in Five Continents (CI5) (Curado et al. 2007). For this reason, incidence rates for adult ovarian GCTs are not presented graphically.

2.3   US Incidence Rates

2.3.1   Pediatric GCTs

Based on data from the National Cancer Institute's Surveillance, Epidemiology and End Results (SEER) program, age-adjusted incidence of pediatric GCTs is 2.8/million in boys and 4.3/million in girls aged 0–14 years in the United States (Horner et al. 2011). The most common histologic subtypes of GCTs in the pediatric age group are teratomas, yolk sac tumors, and germinomas (Fig. 2.6). Teratomas and germinomas had a

higher incidence in girls compared with boys, while there was no significant difference in the incidence of yolk sac tumors.

The higher incidence of pediatric GCTs in girls was observed in all racial and ethnic subgroups, although the difference was not statistically significant for the other category [includes American Indian/Alaskan Native, Asian/Pacific Islander] (Fig. 2.7). The largest difference in incidence between boys and girls was observed in blacks, who have the lowest incidence among boys in this age group (1.7/million) and the highest incidence among girls (5.5/million). These rates were significantly different from the incidence rate in whites in both sexes. No significant differences were observed in Hispanics or children of other races compared with white boys or girls (Fig. 2.7). Previous studies in the USA have reported differences in GCT incidence by race and ethnicity. A recent study of Southeast Asian children in California reported a higher incidence of GCTs in Asians compared with non-Hispanic whites (Ducore et al. 2004). Similarly, previous analyses of the SEER data reported increased incidence of testicular (Poynter et al. 2010a; Walsh et al. 2008) and ovarian (Poynter et al. 2010a) GCTs in Asian/Pacific Islanders and Native Americans compared with whites. These findings are consistent with the high rates of

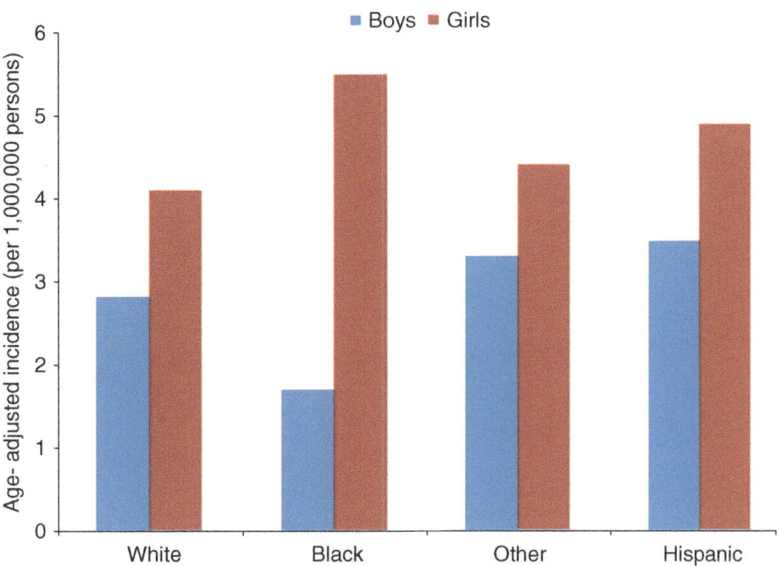

pediatric GCT reported in Asian countries (see Sect. 2.2.1 above). The lack of a significant difference in the data presented here is most likely due to the fact that we did not look at incidence separately for gonadal and extragonadal GCTs, and the increases in previous studies have mainly indicated an excess of gonadal GCTs. A higher incidence of GCTs in Hispanic children in the United States has also been reported (Glazer et al. 1999; Howe et al. 2006; Wilkinson et al. 2005), and one of these studies found that the increased incidence was confined mainly to gonadal GCTs and reached statistical significance only in females (Glazer et al. 1999). Any explanations for these differing incidence patterns would be purely speculative; however, it is possible that genetic factors or differences in hormone levels may play a role.

2.3.2   Adult Testicular GCTs

Incidence rates for adult TGCT in the USA have been reported on extensively using the SEER data (Holmes et al. 2008; McGlynn et al. 2003, 2005; Shah et al. 2007). Adult testicular GCTs are grouped by histology into seminomas and nonseminomas, which includes yolk sac tumors, teratomas, choriocarcinomas, and mixed GCTs

(Sarma et al. 2006). Seminomas are the most common histologic subtype, representing about 60 % of TGCTs (McGlynn et al. 2003). The age-specific incidence of TGCT is slightly different by histologic subtype, with the peak incidence of nonseminomas occurring between the ages of 25–29 years, while the incidence of seminoma peaks between the ages of 35–39 years (Moller 1993). TGCT incidence differs widely by race and ethnic group. Non-Hispanic whites have the highest rates (5.4 per 100,000), followed by Hispanic whites (3.6 per 100,000), Asians and Pacific Islanders/American Indian and Alaskan Natives (2.0 per 100,000), and blacks (0.95 per 100,000) (McGlynn et al. 2003).

2.3.3   Adult Ovarian GCTs

The frequency of GCTs relative to other histologic subtypes of ovarian cancer differs widely in children and adults, with GCTs representing approximately 75 % of malignant ovarian tumors in children (Brookfield et al. 2009) and <5 % of malignant ovarian tumors in adults (dos Santos Silva and Swerdlow 1991; Weiss et al. 1996). Age-adjusted incidence rates over a 30-year period (1973–2002) were 0.34 per 100,000 in the United States (Smith et al. 2006). The highest incidence rates occur in the 15–19-year age range (Poynter et al. 2010a; Smith et al. 2006). Malignant teratomas (39 %) are slightly more common than dysgerminomas (33 %) or mixed/other (29 %) histologic subtypes (Smith et al. 2006). Higher incidence was observed in Asian/Pacific Islanders and Hispanic whites compared with whites; however, these differences only reached borderline statistical significance (Smith et al. 2006).

2.3.4 Trends in Incidence Rates

No clear trend in incidence has been observed in studies of pediatric testicular GCT, with several studies suggesting an increase in incidence (dos Santos Silva et al. 1999; Lacerda et al. 2009; Moller et al. 1995), while others have observed no significant change in incidence (Alanee and Shukla 2009; Swerdlow et al. 1982; Walsh et al. 2006). Using data from the SEER program, no significant increase in incidence was observed in boys ages 0–14 years from 1975 to 2008 (Fig. 2.8). This difference may in part be due to differences in the age distribution of the various study populations. No increase was reported in boys ages 0–4 in studies that looked specifically at this age group (Lacerda et al. 2009; Moller et al. 1995; Poynter et al. 2010a; Walsh et al. 2006), while some studies that included adolescents observed a small increase in incidence (Alanee and Shukla 2009; dos Santos Silva et al. 1999; Dreifaldt et al. 2004; Poynter et al. 2010a).

In contrast to the inconsistent data in children, the increasing incidence of testicular cancer in adults has been well documented (Bergstrom et al. 1996; dos Santos Silva et al 1999; Lacerda et al. 2009; Liu et al. 2000; McGlynn et al. 2003; Richiardi et al. 2004; Weir et al. 1999). This increase is thought to be the result of a birth cohort effect, which supports a role for prenatal exposures in the etiology of this malignancy (Baade et al. 2008; Bergstrom et al. 1996; McGlynn et al. 2003; Richiardi et al. 2004). The birth cohort effect is similar for seminoma and nonseminoma (Bray et al. 2006a, b; Jacobsen et al. 2006; McGlynn et al. 2003), suggesting a common etiology for the two histologic subtypes of TGCT.

Trends in incidence of GCTs in girls have not been studied extensively. Several recent analyses using the SEER data have evaluated ovarian GCTs in both the pediatric and adult populations (Brookfield et al. 2009; Bryant et al. 2009; Kumar et al. 2008; Smith et al. 2006). These data suggest that the incidence of ovarian GCTs has not changed significantly (Smith et al. 2006). Similarly, no significant increase in incidence was reported in girls ages 0–14 years, although the trend line does appear to be increasing slightly (Fig. 2.8).

2.3.5 Survival

Five-year relative survival rates are very high overall for GCTs, mainly due to the effectiveness of platinum-based chemotherapy (Cushing et al. 2004; Einhorn and Donohue 1977; Gobel et al. 2000; Mann et al. 2000). In the United States, 5-year relative survival for the period 1996–2000 was 99 % for boys age 0–14 years, 98 % for girls age 0–14 years, 96 % for males 15–49 years, and 95 % for females 15–49 years (Horner 2011). While survival rates are high overall, differences are observed by tumor characteristics, age at diagnosis, and demographic subgroups. In the pediatric age group where extragonadal tumors are more common (Houldsworth et al. 2006; Rescorla and Breitfeld 1999), survival is higher for tumors located in the gonads than for tumors located in extragonadal locations (Bethel et al. 1998; De Backer et al. 2008; Poynter et al. 2010a; Shah et al. 2008). For ovarian GCTs, survival is higher in the pediatric age group than in the adult age group (Smith et al. 2006). Age appears to play a role in TGCT as well, with men diagnosed after the age of 40 having reduced survival compared with men diagnosed at younger ages (Fossa et al. 2011). Survival differs by race and ethnicity in both males and females in the USA (Biggs and Schwartz 2004; Bryant et al. 2009; Fossa et al. 2011; Smith et al. 2006; Sun et al. 2011).

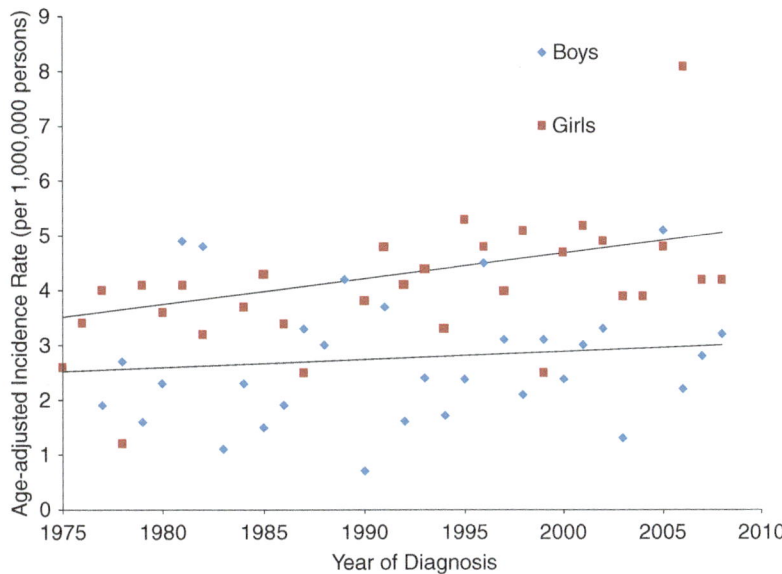

Fig. 2.8 Trends in GCT incidence rates in boys and girls aged 0–14 years in the United States, 1975–2008. Rates are standardized to the 2000 US standard population (*Source*: Horner et al. 2011)

Similar to rates in the USA, incidence of TGCT has declined markedly in most countries (Bray et al. 2006b; Rosen et al. 2011) due to the introduction of platinum-based chemotherapy (Einhorn and Donohue 1977) and standardized recommendations for diagnosis and treatment (Krege et al. 2008). Despite these advances, mortality rates are considerably higher in Central America, Western Asia, and Central and Eastern Europe (Rosen et al. 2011), suggesting a need for improvement in diagnosis and/or treatment in some regions.

2.4 Risk Factors for Pediatric Germ Cell Tumors

The etiology of pediatric germ cell tumors remains largely unknown. Further elucidation of the etiology of GCT in boys may come from results of studies of adult testicular cancer, 95 % of which are GCTs (Schottenfeld 1996). Studies of adult ovarian cancer are less applicable as <5 % are GCTs (dos Santos Silva and Swerdlow 1991; Weiss et al. 1996). To date, few epidemiologic studies have been conducted to evaluate risk factors specifically for pediatric GCTs. Variables that have been evaluated as potential risk factors for

pediatric GCT include family history of cancer, congenital abnormalities, in utero exposure to hormones and pesticides, maternal reproductive history, and parental smoking, occupation, and alcohol consumption, and birth characteristics, and genetic susceptibility (Chen et al. 2005a, b, 2006; Fajardo-Gutierrez et al. 1998; Johnston et al. 1986; Shankar et al. 2006; Shu et al. 1995; Swerdlow et al. 1987; Walker et al. 1988). The evidence for each of these is reviewed briefly below.

2.4.1 Family History

Given the strong heritability of adult testicular GCT (see Sect. 2.5.1), one might hypothesize that family history may also play a role in pediatric GCT. Studies of family history of cancer in children with malignant GCTs are limited in number, with only four previous studies reporting associations between family history of cancer and pediatric GCT (Johnston et al. 1986; Poynter et al. 2010b; Shu et al. 1995; Walker et al. 1988). No clear associations between risk of GCT and family history of cancer emerged, although none of the studies has had power to specifically evaluate the relationship with family history of GCT.

2.4.2 Cryptorchidism and Other Congenital Abnormalities

Cryptorchidism, or undescended testes, is another well-known risk factor for adult TGCT (see Sect. 2.5.2). The largest epidemiologic study to date of pediatric GCTs reported an ~11-fold increased risk of GCTs in males associated with cryptorchidism (Johnson et al. 2009c; Shankar et al. 2006). In this study, the association was observed in males diagnosed both before and after 2 years of age (Johnson et al. 2009c). Cryptorchidism was reported more often than expected in GCT cases in a registry-based study in Japan (Nishi et al. 2000). In addition, cryptorchidism has been reported in several pediatric GCT case series and reports (Huddart et al. 1990; Li and Fraumeni 1972; Mukai et al. 1998). Associations between pediatric GCT and other congenital abnormalities have not been entirely consistent, with associations reported in some (Agha et al. 2005; Johnston et al. 1986, 2009c; Narod et al. 1997; Nishi et al. 2000) but not all (Rankin et al. 2008; Walker et al. 1988) published studies.

2.4.3 Maternal Estrogen Levels

Two case–control studies have evaluated the association between hormonal exposures and malignant pediatric GCTs, with one study focused exclusively on ovarian GCTs (Walker et al. 1988) and the other including all extracranial GCTs in both boys and girls (Shankar et al. 2006). In the ovarian cancer study, the mothers of cases with ovarian GCTs were more likely to have been exposed to hormone drugs during pregnancy than the mothers of controls (OR=3.60, 95 % CI 1.2–13.1 for exposure to hormonal pregnancy test, DES or other supportive hormones, and inadvertent use of oral contraceptives during pregnancy) (Walker et al. 1988). In contrast, no significant association was observed between exogenous hormone use or pregnancy conditions associated with increased hormone levels and pediatric GCT in the study including both boys and girls (Shankar et al. 2006). Similarly, the studies evaluating maternal estrogen levels and adult TGCT have also been mixed (see Sect. 2.5.3). Studies in pediatric GCT are limited by small sample size as well as difficulty in assessing maternal estrogen levels. These conflicting results do not rule out a modest effect of maternal estrogen levels on risk of pediatric GCTs. Further studies with more accurate measures of maternal estrogen levels would be needed to clarify this association.

2.4.4 Parental Exposure to Pesticides and Other Chemicals

Parental exposure to pesticides and other chemicals prior to conception and during gestation, both occupationally and in the home, have been hypothesized as risk factors for childhood cancer, including GCTs. An initial analysis of childhood cancer data collected by the Children's Cancer Group in the United States found that mothers of cases were more likely to report exposure to chemicals and solvents or plastic/resin fumes than mothers of controls, with relatively large risk estimates (OR=4.6, 95 % CI 1.9–11.3 and OR=12.0, 95 % CI 1.9–75.0, respectively) (Shu et al. 1995). Occupational exposure to chemicals was also reported more often in case parents than control parents in a case–control study in the UK (Johnston et al. 1986). However, these results were not confirmed in a larger case–control study by the Children's Oncology Group (Chen et al. 2005b, 2006), although several significant findings were observed in subgroup analyses.

2.4.5 Parental Lifestyle Factors

Given the evidence suggesting that GCTs originate in utero, a number of parental lifestyle exposures have been evaluated as risk factors. Smoking in both mothers (Chen et al. 2005a; Fajardo-Gutierrez et al. 1998; Johnston et al. 1986; Shu et al. 1995) and fathers (Chen et al. 2005a; Fajardo-Gutierrez et al. 1998; Shu et al. 1995) has been evaluated at time points prior to conception (Fajardo-Gutierrez et al. 1998;

Johnston et al. 1986; Shu et al. 1995), during pregnancy (Shu et al. 1995), and after pregnancy (Fajardo-Gutierrez et al. 1998), with most studies showing no association with pediatric GCT. No significant associations have been observed between maternal alcohol consumption (Chen et al. 2005a; Johnston et al. 1986; Shu et al. 1995) and maternal vitamin supplementation (Johnson et al. 2009b; Shu et al. 1995). Maternal dietary patterns during early pregnancy were shown to be associated with GCT in an exploratory analysis from one study (Musselman et al. 2010). Overall, the data do not provide strong evidence that parental lifestyle factors are a significant risk factor for pediatric GCTs, although these characteristics are prone to misclassification and bias in epidemiologic studies, so small associations cannot be ruled out based on the available evidence.

2.4.6 Birth Characteristics

A number of characteristics of the mother and child at birth have been evaluated in association with GCT, including maternal age at the child's birth, birth order, birth weight, birth length, and gestational age. With the exception of one study that reported an increased risk of ovarian GCT in children born to mothers aged less than 20 years (Walker et al. 1988), no significant associations have been reported between maternal (Chen et al. 2005a; Johnston et al. 1986, 2009a; Shu et al. 1995; Stephansson et al. 2011; Wanderas et al. 1998) or paternal age (Johnston et al. 1986, 2009a) and pediatric GCT. Overall, there was no association between birth order and risk of pediatric GCT (Johnston et al. 1986; Shu et al. 1995; Swerdlow et al. 1982; Von Behren et al. 2011; Wanderas et al. 1998); however, one registry-based study showed an increased risk of gonadal GCT among children who were fourth born or later (OR = 1.54, 95 % CI 1.04–2.29)(Von Behren et al. 2011), while another showed a nonsignificant increase in risk of prepubertal testicular GCT for first born compared with later born (OR = 1.40, 95 % CI 0.96–2.05) (Stephansson et al. 2011). Two studies have shown an increased risk with higher birth weight (Chen et al. 2005a; Shu et al.

1995), while another showed no association (Wanderas et al. 1998). Birth length was not significantly associated with prepubertal testicular GCT in two published studies (Stephansson et al. 2011; Wanderas et al. 1998); however, one of these showed an increased risk associated with higher Ponderal index which is measured as the birth weight divided by birth length cubed (kg/m^3) (Stephansson et al. 2011). Three studies reported no association (Chen et al. 2005a; Stephansson et al. 2011; Wanderas et al. 1998) with gestational age, while one reported a significantly lower risk among infants born after 38 weeks of gestation compared with those born at <38 weeks (Shu et al. 1995). With the exception of birth weight, none of the birth characteristics described above has emerged as a potential risk factor for pediatric GCT overall. It is possible that the etiology may differ based on sex or tumor location, but most studies have not been adequately powered to evaluate these subgroup associations.

2.4.7 Genetic Susceptibility

The early age of onset of pediatric cancers suggests that genetic susceptibility may play an important role, through either inherited or de novo mutation. Genetic susceptibility has not been evaluated extensively in pediatric GCTs, although it has been evaluated extensively in adult TGCT (see Sect. 2.5.7). A recent analysis in a small set of pediatric GCTs suggests that some of the susceptibility variants may overlap between pediatric and adult GCTs (Poynter et al. 2012), although a larger study will be required to confirm these findings. Additional studies of genetic susceptibility in pediatric GCTs may increase our understanding of the etiology of these tumors.

2.5 Risk Factors for Adult Testicular Germ Cell Tumors

Risk factors for adult testicular cancer have been discussed in detail elsewhere (Sarma et al. 2006). This chapter will provide a brief review of risk factors that have also been evaluated in the pediatric age group (see Sect. 2.4). Other exposures that have been evaluated as risk factors for TGCT that are not discussed include, but are not limited to, infertility (Raman et al. 2005), adult height and weight (Lerro et al. 2010), and lifestyle factors including smoking, alcohol, physical activity, and diet (Sarma et al. 2006). Risk factors will be discussed for adult testicular cancer overall and will not be stratified by histologic subtype.

2.5.1 Family History

Family history of testicular cancer is one of the few well-established risk factors for adult testicular GCT (Forman et al. 1992), with evidence that the relative risk is much higher in brothers (eight to ten-fold increased risk) than in fathers (four to six-fold increased risk) of testicular GCT patients (Bromen et al. 2004; Heimdal et al. 1996a; Lapes et al. 1977; Polednak 1996; Sonneveld et al. 1999; Westergaard et al. 1996). Among cancers, TGCT has one of the highest reported heritability estimates (Czene et al. 2002), although a positive family history represents only 1–3 % of all cases (Dieckmann and Pichlmeier 1997). This strong association with family history suggests that genetic susceptibility is likely to be important (see Sect. 2.5.7 below). Analyses of family history of other cancers in individuals with testicular GCT have produced less consistent results (Bromen et al. 2004; Chia et al. 2009; Heimdal et al. 1996b; Hemminki and Chen 2006; Hemminki and Li 2004; Kaijser et al. 2003; Kroman et al. 1996; Moss et al. 1986; Spermon et al. 2001; Swerdlow et al. 1987).

2.5.2 Testicular Dysgenesis Syndrome

Cryptorchidism is also a well-known risk factor for adult testicular GCT, with relative risk estimates ranging from 2.5 to 11.4 (Sarma et al. 2006). Based on data from the UK Testicular Cancer Study Group, approximately 10 % of testicular cancers develop in men with cryptorchidism (1994). Cryptorchidism is one abnormality in a spectrum of male reproductive disorders termed testicular dysgenesis syndrome (Boisen et al. 2001; Skakkebaek et al. 2001), of which testicular cancer is the most extreme manifestation (Skakkebaek et al. 2003). This would suggest that cryptorchidism and testicular cancer share common risk factors and that cryptorchidism per se may not be the causal factor. In support of this theory, testicular cancer can occur in the contralateral, normally descended testis in men with unilateral cryptorchidism (Prener et al. 1996); however, the risk is considerably lower compared with risk in the ipsilateral cryptorchid testis (Sarma et al. 2006). Data also suggest that earlier age at orchiopexy reduces risk of subsequent TGCT development, with a sixfold excess in risk in boys who are treated after age 10–11 years or not at all compared with those who are treated earlier (Walsh et al. 2007). These data would support the hypothesis that the microenvironment of the undescended testis contributes to the increased risk of TGCT and that shared environmental or genetic factors for TGCT and cryptorchidism do not provide the entire explanation.

2.5.3 Maternal Estrogen Exposures

Several factors associated with higher estrogen levels in mothers during pregnancy have been associated with testicular cancer in epidemiologic studies, including twin births, birth order, bleeding or spotting during pregnancy, and higher maternal BMI (Cook et al. 2009; Sarma et al. 2006). A recent meta-analysis of perinatal variables in adult TGCT found significant associations for maternal bleeding during pregnancy (OR 1.33, 95 % CI 1.02–1.73), birth order (primiparous vs. not OR = 1.08, 95 % CI 1.01–1.16), and sibship size, while no significant associations were observed for maternal age, maternal nausea, maternal hypertension, preeclampsia, breech delivery, and cesarean section (Cook et al. 2009). These data suggest that maternal estrogen levels may have a modest but significant effect on testicular cancer risk.

In addition to the evidence for endogenous estrogens, some evidence also exists to support a role for exogenous estrogens. Animal studies show that exogenous estrogens lead to the development of components of the testicular dysgenesis syndrome, including cryptorchidism, hypospadias, and dysgenetic gonads (Bullock et al. 1988; McLachlan et al. 1975). Exogenous estrogen use in early pregnancy has also been associated with an increased risk of TGCT in epidemiologic studies (Henderson et al. 1979; Schottenfeld et al. 1980; Weir et al. 2000), although analyses of diethylstilbestrol (DES) exposure, a potent synthetic estrogen, have shown mixed results (Giusti et al. 1995).

2.5.4 Pesticides and Other Endocrine-Disrupting Agents

Pesticide exposure has been hypothesized to increase risk of testicular cancer. The incidence of testicular cancer was higher than expected in an analysis of the offspring of British agricultural pesticide users (Frost et al. 2011), while a similar study in Sweden did not find an increased incidence (Rodvall et al. 2003). Endocrine-disrupting agents in particular have been hypothesized to increase risk due to their ability to act as weak estrogens or antiandrogens by binding to the estrogen and androgen receptors (McGlynn 2001). Data from in vitro and in vivo studies provides evidence that exposure to estrogens and endocrine disruptors may influence germ cell apoptosis (Chaki et al. 2006; Delbes et al. 2004; Lambrot et al. 2009; Li et al. 2009) and stimulate cell proliferation (Bouskine et al. 2008, 2009). Persistent organochlorine pesticides (POP) (Biggs et al. 2008; Hardell et al. 2003; McGlynn et al. 2008) and polychlorinated biphenyls (Hardell et al. 2003; McGlynn et al. 2009) have been investigated in epidemiologic studies of adult TGCT, with evidence suggesting that POPs (McGlynn et al. 2008) may be associated with increased risk.

2.5.5 Maternal Lifestyle Factors During Pregnancy

Maternal smoking and alcohol consumption have also been evaluated as risk factors for adult TGCT. Similar to studies in pediatric GCT (see Sect. 2.4.5 above), the data do not support a role for either maternal smoking (Moller and Skakkebaek 1997; Moller and Westergaard 1998; Weir et al. 2000) or maternal alcohol consumption (Brown et al. 1986; Weir et al. 2000) during pregnancy and subsequent risk of TGCT in the offspring.

2.5.6 Birth Characteristics

Because testicular cancer is hypothesized to be initiated in utero, numerous studies have evaluated the association between birth characteristics and TGCT, including birth weight and length, birth order, gestational age, twin birth, breastfeeding, and neonatal jaundice. Of these, a recent meta-analysis supports an association between twin birth (OR = 1.22, 95 % CI 1.03–1.44), birth weight (OR = 1.34, 95 % CI 1.08–1.67 for low vs. not), and gestational age (OR = 0.95, 95 % CI 0.92–0.98 per week) and TGCT (Cook et al. 2010).

2.5.7 Genetic Susceptibility

Given the strong heritability observed in adult TGCT (Hemminki and Li 2004), genetic susceptibility has been studied extensively (Rapley 2007). A genome-wide linkage scan for testicular cancer identified no significant susceptibility alleles, and the authors concluded that the genetic risk was most likely due to multiple variants with small effects rather that one major susceptibility gene (Crockford et al. 2006). Consistent with this hypothesis, a number of potential susceptibility alleles have been identified in recent years using both candidate gene studies and GWAS. Candidate gene studies have suggested that variation in genes involved in catechol estrogen metabolism (Figueroa et al. 2008; Starr et al. 2005), immune function (Purdue et al. 2007), the inhibin pathway (Purdue et al. 2008), and a Y chromosome gr/gr deletion (Nathanson et al. 2005) may be associated with adult TGCT risk. In addition, the gr/gr deletion (Nathanson et al. 2005) and mutations in *PDE11A* (Horvath et al. 2009) may be especially relevant in familial TGCT. Recent GWAS identified susceptibility loci for TGCT near *KITLG*, *SPRY4*, and *BAK1* (Kanetsky et al. 2009; Rapley et al. 2009). Secondary analyses of GWAS data uncovered additional susceptibility loci near *DMRT1*, *TERT*, and *ATF7IP* (Kanetsky et al. 2011; Turnbull et al. 2010). Several of these loci have been confirmed in a subsequent analysis of familial and bilateral TGCT (Kratz et al. 2011). These loci were notable both for their relatively large effect sizes compared to GWAS of other cancers and because they were located in regions with genes highly relevant to germ cell biology. Additional fine mapping and functional studies will help elucidate the potential roles of these genes in GCTs.

2.6 Risk Factors for Adult Ovarian Germ Cell Tumors

Given the low incidence of malignant ovarian GCTs in adult women (Fig. 2.2), few studies have been conducted to evaluate the etiology of these tumors. Family history has been evaluated in the literature, with mixed results. A study of 74 cases of ovarian GCT found none with a family history (Shulman et al. 1994), while other reports have described families with cases of both ovarian GCT and either TGCT or extragonadal GCT (Giambartolomei et al. 2009). The few risk factors that have been evaluated include inflammatory cytokines during pregnancy (Toriola et al. 2011) and maternal hormone exposure in cases diagnosed before age 35 years (Walker et al. 1988). The rare nature of GCTs in older women limits our ability to identify risk factors. Pooled analysis of existing datasets may provide a sufficient sample size for analysis.

2.7 Summary and Conclusions

Evidence suggests that GCTs, including those in adults, are initiated in utero. Thus, alterations in normal embryonic development are likely to be especially relevant to GCT etiology. Studies in mouse models that resemble pediatric GCTs and studies of adult TGCT have identified susceptibility alleles in genes essential for the normal development of PGCs, which deserves further investigation. Several variants identified in GWAS of adult TGCT have also been shown to be associated with GCT in the pediatric age group, including in girls (Poynter et al 2012). This suggests that at least some GCT risk factors are common to all age groups.

In addition to shared genetic susceptibility, recent studies also suggest that cryptorchidism and possibly birth weight are shared risk factors for pediatric and adult testicular GCT (see Sects. 2.4 and 2.5). In contrast, other risk factors for TGCT in adults, including maternal estrogen exposure, have not been associated with GCT in the pediatric age group. Given the evidence that adult testicular GCTs arise from a more mature primordial germ cell (Oosterhuis and Looijenga 2005), it is not surprising that the etiology is not completely overlapping.

Less is known about risk for GCTs in females, especially in the adult age group. The extremely rare occurrence of malignant GCT in the adult ovary has made it especially difficult to understand etiology. Pooled analysis of available datasets may help elucidate potential risk factors. In contrast to malignant ovarian GCTs, benign ovarian teratomas are more common in adult women (Westhoff et al. 1988). Comparisons of benign and malignant ovarian teratomas may also provide some clues.

Recent advances in genomic technology will provide exciting opportunities to better understand the etiology of this rare and understudied group of tumors. In addition to providing valuable information regarding etiology, studies of GCTs may have far-reaching relevance given the recent interest in the role of stem cells in carcinogenesis (Clark 2007; Visvader and Lindeman 2008). Since early epigenetic reprogramming restores pluripotency in germ cells (Donovan and de Miguel 2003), a better understanding of GCTs could provide important insights into the fetal origins of carcinogenesis and contribute insights into treatment and prevention of pediatric and adult tumors.

Acknowledgments This work was supported in part by the Children's Cancer Research Fund, Minneapolis, MN.

References

Agha MM, Williams JI, Marrett L, To T, Zipursky A, Dodds L (2005) Congenital abnormalities and childhood cancer. Cancer 103:1939–1948

Alanee S, Shukla A (2009) Paediatric testicular cancer: an updated review of incidence and conditional survival from the Surveillance, Epidemiology and End Results database. BJU Int 104:1280–1283

Baade P, Carriere P, Fritschi L (2008) Trends in testicular germ cell cancer incidence in Australia. Cancer Causes Control 19:1043–1049

Bergstrom R, Adami HO, Mohner M, Zatonski W, Storm H, Ekbom A, Tretli S, Teppo L, Akre O, Hakulinen T (1996) Increase in testicular cancer incidence in six European countries: a birth cohort phenomenon. J Natl Cancer Inst 88:727–733

Bernstein L, Smith MA, Liu L, Deapen D, Friedman DL (1999) Germ cell, trophoblastic, and other gonadal neoplasms. In: Ries L, Smith MA, Gurney JG, Linet M, Tamra T, Young TL, Bunin GR (eds) Cancer incidence and survival among children and adolescents: United States SEER Program 1975-1995. NIH Pub. No. 99-4649. National Cancer Institute, SEER Program, Bethesda, pp 125–137

Bethel CA, Mutabagani K, Hammond S, Besner GE, Caniano DA, Cooney DR (1998) Nonteratomatous germ cell tumors in children. J Pediatr Surg 33:1122–1126; discussion 1126–1127

Biggs ML, Davis MD, Eaton DL, Weiss NS, Barr DB, Doody DR, Fish S, Needham LL, Chen C, Schwartz SM (2008) Serum organochlorine pesticide residues and risk of testicular germ cell carcinoma: a population-based case–control study. Cancer Epidemiol Biomarkers Prev 17:2012–2018

Biggs ML, Schwartz SM (2004) Differences in testis cancer survival by race and ethnicity: a population-based study, 1973–1999 (United States). Cancer Causes Control 15:437–444

Boisen KA, Main KM, Rajpert-De Meyts E, Skakkebaek NE (2001) Are male reproductive disorders a common entity? The testicular dysgenesis syndrome. Ann N Y Acad Sci 948:90–99

Bouskine A, Nebout M, Brucker-Davis F, Benahmed M, Fenichel P (2009) Low doses of bisphenol A promote human seminoma cell proliferation by activating PKA and PKG via a membrane G-protein-coupled estrogen receptor. Environ Health Perspect 117:1053–1058

Bouskine A, Nebout M, Mograbi B, Brucker-Davis F, Roger C, Fenichel P (2008) Estrogens promote human testicular germ cell cancer through a membrane-mediated activation of extracellular regulated kinase and protein kinase A. Endocrinology 149:565–573

Bray F, Richiardi L, Ekbom A, Forman D, Pukkala E, Cuninkova M, Moller H (2006a) Do testicular seminoma and nonseminoma share the same etiology? Evidence from an age-period-cohort analysis of incidence trends in eight European countries. Cancer Epidemiol Biomarkers Prev 15:652–658

Bray F, Richiardi L, Ekbom A, Pukkala E, Cuninkova M, Moller H (2006b) Trends in testicular cancer incidence and mortality in 22 European countries: continuing increases in incidence and declines in mortality. Int J Cancer 118:3099–3111

Bromen K, Stang A, Baumgardt-Elms C, Stegmaier C, Ahrens W, Metz KA, Jockel KH (2004) Testicular, other genital, and breast cancers in first-degree relatives of testicular cancer patients and controls. Cancer Epidemiol Biomarkers Prev 13:1316–1324

Brookfield KF, Cheung MC, Koniaris LG, Sola JE, Fischer AC (2009) A population-based analysis of 1037 malignant ovarian tumors in the pediatric population. J Surg Res 156:45–49

Brown LM, Pottern LM, Hoover RN (1986) Prenatal and perinatal risk factors for testicular cancer. Cancer Res 46:4812–4816

Bryant CS, Kumar S, Shah JP, Mahdi H, Ali-Fehmi R, Munkarah AR, Deppe G, Morris RT (2009) Racial disparities in survival among patients with germ cell tumors of the ovary – United States. Gynecol Oncol 114:437–441

Bullock BC, Newbold RR, McLachlan JA (1988) Lesions of testis and epididymis associated with prenatal diethylstilbestrol exposure. Environ Health Perspect 77:29–31

Chaki SP, Misro MM, Gautam DK, Kaushik M, Ghosh D, Chainy GB (2006) Estradiol treatment induces testicular oxidative stress and germ cell apoptosis in rats. Apoptosis 11:1427–1437

Chen Z, Robison L, Giller R, Krailo M, Davis M, Davies S, Shu XO (2006) Environmental exposure to residential pesticides, chemicals, dusts, fumes, and metals, and risk of childhood germ cell tumors. Int J Hyg Environ Health 209:31–40

Chen Z, Robison L, Giller R, Krailo M, Davis M, Gardner K, Davies S, Shu XO (2005a) Risk of childhood germ cell tumors in association with parental smoking and drinking. Cancer 103:1064–1071

Chen Z, Stewart PA, Davies S, Giller R, Krailo M, Davis M, Robison L, Shu XO (2005b) Parental occupational exposure to pesticides and childhood germ-cell tumors. Am J Epidemiol 162:858–867

Chia VM, Li Y, Goldin LR, Graubard BI, Greene MH, Korde L, Rubertone MV, Erickson RL, McGlynn KA (2009) Risk of cancer in first- and second-degree relatives of testicular germ cell tumor cases and controls. Int J Cancer 124:952–957

Chia VM, Quraishi SM, Devesa SS, Purdue MP, Cook MB, McGlynn KA (2010) International trends in the incidence of testicular cancer, 1973–2002. Cancer Epidemiol Biomarkers Prev 19:1151–1159

Clark AT (2007) The stem cell identity of testicular cancer. Stem Cell Rev 3:49–59

Cook MB, Akre O, Forman D, Madigan MP, Richiardi L, McGlynn KA (2009) A systematic review and meta-analysis of perinatal variables in relation to the risk of testicular cancer – experiences of the mother. Int J Epidemiol 38:1532–1542

Cook MB, Akre O, Forman D, Madigan MP, Richiardi L, McGlynn KA (2010) A systematic review and meta-analysis of perinatal variables in relation to the risk of testicular cancer – experiences of the son. Int J Epidemiol 39:1605–1618

Crockford GP, Linger R, Hockley S, Dudakia D, Johnson L, Huddart R, Tucker K, Friedlander M, Phillips KA, Hogg D, Jewett MA, Lohynska R, Daugaard G, Richard S, Chompret A, Bonaiti-Pellie C, Heidenreich A, Albers P, Olah E, Geczi L, Bodrogi I, Ormiston WJ, Daly PA, Guilford P, Fossa SD, Heimdal K, Tjulandin SA, Liubchenko L, Stoll H, Weber W, Forman D, Oliver T, Einhorn L, McMaster M, Kramer J, Greene MH, Weber BL, Nathanson KL, Cortessis V, Easton DF, Bishop DT, Stratton MR, Rapley EA (2006) Genome-wide linkage screen for testicular germ cell tumour susceptibility loci. Hum Mol Genet 15:443–451

Curado MP, Edwards B, Shin HR, Storm H, Ferlay J, Heanue M, Boyle P (eds) (2007) Cancer incidence in five continents, vol IX. IARC, Lyon

Cushing B, Giller R, Cullen JW, Marina NM, Lauer SJ, Olson TA, Rogers PC, Colombani P, Rescorla F, Billmire DF, Vinocur CD, Hawkins EP, Davis MM, Perlman EJ, London WB, Castleberry RP (2004) Randomized comparison of combination chemotherapy with etoposide, bleomycin, and either high-dose or standard-dose cisplatin in children and adolescents with high-risk malignant germ cell tumors: a pediatric intergroup study – Pediatric Oncology Group 9049 and Children's Cancer Group 888. J Clin Oncol 22:2691–2700

Czene K, Lichtenstein P, Hemminki K (2002) Environmental and heritable causes of cancer among 9.6 million individuals in the Swedish Family-Cancer Database. Int J Cancer 99:260–266

De Backer A, Madern GC, Pieters R, Haentjens P, Hakvoort-Cammel FG, Oosterhuis JW, Hazebroek FW (2008) Influence of tumor site and histology on long-term survival in 193 children with extracranial germ cell tumors. Eur J Pediatr Surg 18:1–6

Delbes G, Levacher C, Pairault C, Racine C, Duquenne C, Krust A, Habert R (2004) Estrogen receptor beta-mediated inhibition of male germ cell line development in mice by endogenous estrogens during perinatal life. Endocrinology 145:3395–3403

Dieckmann KP, Pichlmeier U (1997) The prevalence of familial testicular cancer: an analysis of two patient populations and a review of the literature. Cancer 80:1954–1960

Donovan PJ, de Miguel MP (2003) Turning germ cells into stem cells. Curr Opin Genet Dev 13:463–471

dos Santos Silva I, Swerdlow AJ (1991) Ovarian germ cell malignancies in England: epidemiological parallels with testicular cancer. Br J Cancer 63:814–818

dos Santos Silva I, Swerdlow AJ, Stiller CA, Reid A (1999) Incidence of testicular germ-cell malignancies in England and Wales: trends in children compared with adults. Int J Cancer 83:630–634

Dreifaldt AC, Carlberg M, Hardell L (2004) Increasing incidence rates of childhood malignant diseases in Sweden during the period 1960–1998. Eur J Cancer 40:1351–1360

Ducore JM, Parikh-Patel A, Gold EB (2004) Cancer occurrence in Southeast Asian children in California. J Pediatr Hematol Oncol 26:613–618

Einhorn LH, Donohue J (1977) Cis-diamminedichloroplatinum, vinblastine, and bleomycin combination chemotherapy in disseminated testicular cancer. Ann Intern Med 87:293–298

Fajardo-Gutierrez A, Gomez-Gomez M, Danglot-Banck C, Alvarez-Contreras JJ, Yamamoto-Kimura L (1998) Risk factors for the development of germ cell tumors in children. Gac Med Mex 134:273–281

Figueroa JD, Sakoda LC, Graubard BI, Chanock S, Rubertone MV, Erickson RL, McGlynn KA (2008) Genetic variation in hormone metabolizing genes and risk of testicular germ cell tumors. Cancer Causes Control 19:917–929

Forman D, Oliver RT, Brett AR, Marsh SG, Moses JH, Bodmer JG, Chilvers CE, Pike MC (1992) Familial testicular cancer: a report of the UK family register, estimation of risk and an HLA class 1 sib-pair analysis. Br J Cancer 65:255–262

Fossa SD, Cvancarova M, Chen L, Allan AL, Oldenburg J, Peterson DR, Travis LB (2011) Adverse prognostic factors for testicular cancer-specific survival: a population-based study of 27,948 patients. J Clin Oncol 29:963–970

Frost G, Brown T, Harding AH (2011) Mortality and cancer incidence among British agricultural pesticide users. Occup Med (Lond) 61:303–310

Giambartolomei C, Mueller CM, Greene MH, Korde LA (2009) A mini-review of familial ovarian germ cell tumors: an additional manifestation of the familial testicular germ cell tumor syndrome. Cancer Epidemiol 33:31–36

Giusti RM, Iwamoto K, Hatch EE (1995) Diethylstilbestrol revisited: a review of the long-term health effects. Ann Intern Med 122:778–788

Glazer ER, Perkins CI, Young JL Jr, Schlag RD, Campleman SL, Wright WE (1999) Cancer among Hispanic children in California, 1988–1994: comparison with non-Hispanic white children. Cancer 86:1070–1079

Gobel U, Schneider DT, Calaminus G, Haas RJ, Schmidt P, Harms D (2000) Germ-cell tumors in childhood and adolescence. GPOH MAKEI and the MAHO study groups. Ann Oncol 11:263–271

Hardell L, van Bavel B, Lindstrom G, Carlberg M, Dreifaldt AC, Wijkstrom H, Starkhammar H, Eriksson M, Hallquist A, Kolmert T (2003) Increased concentrations of polychlorinated biphenyls, hexachlorobenzene, and chlordanes in mothers of men with testicular cancer. Environ Health Perspect 111:930–934

Harms D, Janig U (1986) Germ cell tumours of childhood. Report of 170 cases including 59 pure and partial yolk-sac tumours. Virchows Arch A Pathol Anat Histopathol 409:223–239

Heimdal K, Olsson H, Tretli S, Flodgren P, Borresen AL, Fossa SD (1996a) Familial testicular cancer in Norway and southern Sweden. Br J Cancer 73:964–969

Heimdal K, Olsson H, Tretli S, Flodgren P, Borresen AL, Fossa SD (1996b) Risk of cancer in relatives of testicular cancer patients. Br J Cancer 73:970–973

Hemminki K, Chen B (2006) Familial risks in testicular cancer as aetiological clues. Int J Androl 29:205–210

Hemminki K, Li X (2004) Familial risk in testicular cancer as a clue to a heritable and environmental aetiology. Br J Cancer 90:1765–1770

Henderson BE, Benton B, Jing J, Yu MC, Pike MC (1979) Risk factors for cancer of the testis in young men. Int J Cancer 23:598–602

Holmes L Jr, Escalante C, Garrison O, Foldi BX, Ogungbade GO, Essien EJ, Ward D (2008) Testicular cancer incidence trends in the USA (1975–2004): plateau or shifting racial paradigm? Public Health 122:862–872

Horner MJ RL, Krapcho M, Neyman N, Aminou R, Howlader N, Altekruse SF, Feuer EJ, Huang L, Mariotto A, Miller BA, Lewis DR, Eisner MP, Stinchcomb DG, Edwards BK (eds) (2011) SEER cancer statistics review, 1975 2008. National Cancer Institute, Bethesda (Based on November 2010 SEER data submission, posted to the SEER web site, 2011)

Horvath A, Korde L, Greene MH, Libe R, Osorio P, Faucz FR, Raffin-Sanson ML, Tsang KM, Drori-Herishanu L, Patronas Y, Remmers EF, Nikita ME, Moran J, Greene J, Nesterova M, Merino M, Bertherat J, Stratakis CA (2009) Functional phosphodiesterase 11A mutations may modify the risk of familial and bilateral testicular germ cell tumors. Cancer Res 69:5301–5306

Houldsworth J, Korkola JE, Bosl GJ, Chaganti RS (2006) Biology and genetics of adult male germ cell tumors. J Clin Oncol 24:5512–5518

Howe HL, Wu X, Ries LA, Cokkinides V, Ahmed F, Jemal A, Miller B, Williams M, Ward E, Wingo PA, Ramirez A, Edwards BK (2006) Annual report to the nation on the status of cancer, 1975–2003, featuring cancer among U.S. Hispanic/Latino populations. Cancer 107:1711–1742

Huddart SN, Mann JR, Gornall P, Pearson D, Barrett A, Raafat F, Barnes JM, Wallendsus KR (1990) The UK Children's Cancer Study Group: testicular malignant germ cell tumours 1979–1988. J Pediatr Surg 25:406–410

Jacobsen R, Moller H, Thoresen SO, Pukkala E, Kjaer SK, Johansen C (2006) Trends in testicular cancer incidence in the Nordic countries, focusing on the recent decrease in Denmark. Int J Androl 29:199–204

Johnson KJ, Carozza SE, Chow EJ, Fox EE, Horel S, McLaughlin CC, Mueller BA, Puumala SE, Reynolds P, Von Behren J, Spector LG (2009a) Parental age and risk of childhood cancer: a pooled analysis. Epidemiology 20:475–483

Johnson KJ, Poynter JN, Ross JA, Robison LL, Shu XO (2009b) Pediatric germ cell tumors and maternal vitamin supplementation: a Children's Oncology Group study. Cancer Epidemiol Biomarkers Prev 18:2661–2664

Johnson KJ, Ross JA, Poynter JN, Linabery AM, Robison LL, Shu XO (2009c) Paediatric germ cell tumours and congenital abnormalities: a Children's Oncology Group study. Br J Cancer 101:518–521

Johnston HE, Mann JR, Williams J, Waterhouse JA, Birch JM, Cartwright RA, Draper GJ, Hartley AL, McKinney PA, Hopton PA et al (1986) The Inter-Regional, Epidemiological Study of Childhood Cancer (IRESCC): case–control study in children with germ cell tumours. Carcinogenesis 7:717–722

Kaijser M, Akre O, Cnattingius S, Ekbom A (2003) Maternal lung cancer and testicular cancer risk in the offspring. Cancer Epidemiol Biomarkers Prev 12:643–646

Kanetsky PA, Mitra N, Vardhanabhuti S, Li M, Vaughn DJ, Letrero R, Ciosek SL, Doody DR, Smith LM, Weaver J, Albano A, Chen C, Starr JR, Rader DJ, Godwin AK, Reilly MP, Hakonarson H, Schwartz SM, Nathanson KL (2009) Common variation in KITLG and at 5q31.3 predisposes to testicular germ cell cancer. Nat Genet 41:811–815

Kanetsky PA, Mitra N, Vardhanabhuti S, Vaughn DJ, Li M, Ciosek SL, Letrero R, D'Andrea K, Vaddi M, Doody DR, Weaver J, Chen C, Starr JR, Hakonarson H, Rader DJ, Godwin AK, Reilly MP, Schwartz SM, Nathanson KL (2011) A second independent locus within DMRT1 is associated with testicular germ cell tumor susceptibility. Hum Mol Genet 20:3109–3117

Kratz CP, Han SS, Rosenberg PS, Berndt SI, Burdett L, Yeager M, Korde LA, Mai PL, Pfeiffer R, Greene MH (2011) Variants in or near KITLG, BAK1, DMRT1, and TERT-CLPTM1L predispose to familial testicular germ cell tumour. J Med Genet 48:473–476

Krege S, Beyer J, Souchon R, Albers P, Albrecht W, Algaba F, Bamberg M, Bodrogi I, Bokemeyer C, Cavallin-Stahl E, Classen J, Clemm C, Cohn-Cedermark G, Culine S, Daugaard G, De Mulder PH, De Santis M, de Wit M, de Wit R, Derigs HG, Dieckmann KP, Dieing A, Droz JP, Fenner M, Fizazi K, Flechon A, Fossa SD, del Muro XG, Gauler T, Geczi L, Gerl A, Germa-Lluch JR, Gillessen S, Hartmann JT, Hartmann M, Heidenreich A, Hoeltl W, Horwich A, Huddart R, Jewett M, Joffe J, Jones WG, Kisbenedek L, Klepp O, Kliesch S, Koehrmann KU, Kollmannsberger C, Kuczyk M, Laguna P, Galvis OL, Loy V, Mason MD, Mead GM, Mueller R, Nichols C, Nicolai N, Oliver T, Ondrus D, Oosterhof GO, Ares LP, Pizzocaro G, Pont J, Pottek T, Powles T, Rick O, Rosti G, Salvioni R, Scheiderbauer J, Schmelz HU, Schmidberger H, Schmoll HJ, Schrader M, Sedlmayer F, Skakkebaek NE, Sohaib A, Tjulandin S, Warde P, Weinknecht S, Weissbach L, Wittekind C, Winter E, Wood L, von der Maase H (2008) European consensus conference on diagnosis and treatment of germ cell cancer: a report of the second meeting of the European Germ Cell Cancer Consensus group (EGCCCG): part I. Eur Urol 53:478–496

Kroman N, Frisch M, Olsen JH, Westergaard T, Melbye M (1996) Oestrogen-related cancer risk in mothers of testicular-cancer patients. Int J Cancer 66:438–440

Kumar S, Shah JP, Bryant CS, Imudia AN, Cote ML, Ali-Fehmi R, Malone JM Jr, Morris RT (2008) The prevalence and prognostic impact of lymph node metastasis in malignant germ cell tumors of the ovary. Gynecol Oncol 110:125–132

Kuratsu J, Ushio Y (1996) Epidemiological study of primary intracranial tumors in childhood. A population-based survey in Kumamoto Prefecture, Japan. Pediatr Neurosurg 25:240–246; discussion 247

Lacerda HM, Akre O, Merletti F, Richiardi L (2009) Time trends in the incidence of testicular cancer in childhood and young adulthood. Cancer Epidemiol Biomarkers Prev 18:2042–2045

Lambrot R, Muczynski V, Lecureuil C, Angenard G, Coffigny H, Pairault C, Moison D, Frydman R, Habert R, Rouiller-Fabre V (2009) Phthalates impair germ cell development in the human fetal testis in vitro without change in testosterone production. Environ Health Perspect 117:32–37

Lapes M, Iozzi L, Ziegenfus WD, Antoniades K, Vivacqua R (1977) Familial testicular cancer in a father (bilateral seminoma-embryonal cell carcinoma) and son (teratocarcinoma): a case report and review of the literature. Cancer 39:2317–2320

Lerro CC, McGlynn KA, Cook MB (2010) A systematic review and meta-analysis of the relationship between body size and testicular cancer. Br J Cancer 103:1467–1474

Li FP, Fraumeni JF (1972) Testicular cancers in children: epidemiologic characteristics. J Natl Cancer Inst 48:1575–1581

Li YJ, Song TB, Cai YY, Zhou JS, Song X, Zhao X, Wu XL (2009) Bisphenol A exposure induces apoptosis and upregulation of Fas/FasL and caspase-3 expression in the testes of mice. Toxicol Sci 108:427–436

Liu S, Semenciw R, Waters C, Wen SW, Mery LS, Mao Y (2000) Clues to the aetiological heterogeneity of testicular seminomas and non-seminomas: time trends and age-period-cohort effects. Int J Epidemiol 29:826–831

Mann JR, Raafat F, Robinson K, Imeson J, Gornall P, Sokal M, Gray E, McKeever P, Hale J, Bailey S, Oakhill A (2000) The United Kingdom Children's Cancer Study Group's second germ cell tumor study: carboplatin, etoposide, and bleomycin are effective treatment for children with malignant extracranial germ cell tumors, with acceptable toxicity. J Clin Oncol 18:3809–3818

McGlynn KA (2001) Environmental and host factors in testicular germ cell tumors. Cancer Invest 19:842–853

McGlynn KA, Devesa SS, Graubard BI, Castle PE (2005) Increasing incidence of testicular germ cell tumors among black men in the United States. J Clin Oncol 23:5757–5761

McGlynn KA, Devesa SS, Sigurdson AJ, Brown LM, Tsao L, Tarone RE (2003) Trends in the incidence of testicular germ cell tumors in the United States. Cancer 97:63–70

McGlynn KA, Quraishi SM, Graubard BI, Weber JP, Rubertone MV, Erickson RL (2008) Persistent organochlorine pesticides and risk of testicular germ cell tumors. J Natl Cancer Inst 100:663–671

McGlynn KA, Quraishi SM, Graubard BI, Weber JP, Rubertone MV, Erickson RL (2009) Polychlorinated biphenyls and risk of testicular germ cell tumors. Cancer Res 69:1901–1909

McLachlan JA, Newbold RR, Bullock B (1975) Reproductive tract lesions in male mice exposed prenatally to diethylstilbestrol. Science 190:991–992

Moller H (1993) Clues to the aetiology of testicular germ cell tumours from descriptive epidemiology. Eur Urol 23:8–13; discussion 14–15

Moller H, Evans H (2003) Epidemiology of gonadal germ cell cancer in males and females. Apmis 111:43–46; discussion 46–48

Moller H, Jorgensen N, Forman D (1995) Trends in incidence of testicular cancer in boys and adolescent men. Int J Cancer 61:761–764

Moller H, Skakkebaek NE (1997) Testicular cancer and cryptorchidism in relation to prenatal factors: case–control studies in Denmark. Cancer Causes Control 8:904–912

Moller H, Westergaard T (1998) Tobacco smoking and testicular cancer. Ugeskr Laeger 160:1041–1042

Moss AR, Osmond D, Bacchetti P, Torti FM, Gurgin V (1986) Hormonal risk factors in testicular cancer. A case–control study. Am J Epidemiol 124:39–52

Motta PM, Nottola SA, Makabe S (1997) Natural history of the female germ cell from its origin to full maturation through prenatal ovarian development. Eur J Obstet Gynecol Reprod Biol 75:5–10

Mukai M, Takamatsu H, Noguchi H, Tahara H (1998) Intra-abdominal testis with mature teratoma. Pediatr Surg Int 13:204–205

Musselman JR, Jurek AM, Johnson KJ, Linabery AM, Robison LL, Shu XO, Ross JA (2010) Maternal dietary patterns during early pregnancy and the odds of childhood germ cell tumors: a Children's Oncology Group study. Am J Epidemiol 173:282–291

Narod SA, Hawkins MM, Robertson CM, Stiller CA (1997) Congenital anomalies and childhood cancer in Great Britain. Am J Hum Genet 60:474–485

Nathanson KL, Kanetsky PA, Hawes R, Vaughn DJ, Letrero R, Tucker K, Friedlander M, Phillips KA, Hogg D, Jewett MA, Lohynska R, Daugaard G, Richard S, Chompret A, Bonaiti-Pellie C, Heidenreich A, Olah E, Geczi L, Bodrogi I, Ormiston WJ, Daly PA, Oosterhuis JW, Gillis AJ, Looijenga LH, Guilford P, Fossa SD, Heimdal K, Tjulandin SA, Liubchenko L, Stoll H, Weber W, Rudd M, Huddart R, Crockford GP, Forman D, Oliver DT, Einhorn L, Weber BL, Kramer J, McMaster M, Greene MH, Pike M, Cortessis V, Chen C, Schwartz SM, Bishop DT, Easton DF, Stratton MR, Rapley EA (2005) The Y deletion gr/gr and susceptibility to testicular germ cell tumor. Am J Hum Genet 77:1034–1043

Nicolaides NC, Papadopoulos N, Liu B, Wei YF, Carter KC, Ruben SM, Rosen CA, Haseltine WA, Fleischmann RD, Fraser CM et al (1994) Mutations of two PMS homologues in hereditary nonpolyposis colon cancer. Nature 371:75–80

Nishi M, Miyake H, Takeda T, Hatae Y (2000) Congenital malformations and childhood cancer. Med Pediatr Oncol 34:250–254

Oosterhuis JW, Looijenga LH (2005) Testicular germ-cell tumours in a broader perspective. Nat Rev Cancer 5:210–222

Oosterhuis JW, Stoop H, Honecker F, Looijenga LH (2007) Why human extragonadal germ cell tumours occur in the midline of the body: old concepts, new perspectives. Int J Androl 30:256–263; discussion 263–264

Parkin DM, Kramarova E, Draper GJ, Masuyer E, Michaelis J, Neglia J, Qureshi S, Stiller CA (eds) (1998) International incidence of childhood cancer. International Agency for Research on Cancer/World Health Organization, Lyon

Polednak AP (1996) Familial testicular cancer in a population-based cancer registry. Urol Int 56:238–240

Poynter JN, Amatruda JF, Ross JA (2010a) Trends in incidence and survival of pediatric and adolescent patients with germ cell tumors in the United States, 1975 to 2006. Cancer 116:4882–4891

Poynter JN, Hooten AJ, Frazier AL, Ross JA (2012) Associations between variants in KITLG, SPRY4, BAK1 and DMRT1 and pediatric germ cell tumors. Genes Chromosomes Cancer 51:266–271

Poynter JN, Radzom AH, Spector LG, Puumala S, Robison LL, Chen Z, Ross JA, Shu XO (2010b) Family history of cancer and malignant germ cell tumors in children: a report from the Children's Oncology Group. Cancer Causes Control 21:181–189

Prener A, Engholm G, Jensen OM (1996) Genital anomalies and risk for testicular cancer in Danish men. Epidemiology 7:14–19

Purdue MP, Devesa SS, Sigurdson AJ, McGlynn KA (2005) International patterns and trends in testis cancer incidence. Int J Cancer 115:822–827

Purdue MP, Graubard BI, Chanock SJ, Rubertone MV, Erickson RL, McGlynn KA (2008) Genetic variation in the inhibin pathway and risk of testicular germ cell tumors. Cancer Res 68:3043–3048

Purdue MP, Sakoda LC, Graubard BI, Welch R, Chanock SJ, Sesterhenn IA, Rubertone MV, Erickson RL, McGlynn KA (2007) A case–control investigation of immune function gene polymorphisms and risk of testicular germ cell tumors. Cancer Epidemiol Biomarkers Prev 16:77–83

Raman JD, Nobert CF, Goldstein M (2005) Increased incidence of testicular cancer in men presenting with infertility and abnormal semen analysis. J Urol 174:1819–1822; discussion 1822

Rankin J, Silf KA, Pearce MS, Parker L, Ward Platt M (2008) Congenital anomaly and childhood cancer: a population-based, record linkage study. Pediatr Blood Cancer 51:608–612

Rapley E (2007) Susceptibility alleles for testicular germ cell tumour: a review. Int J Androl 30:242–250; discussion 250

Rapley EA, Turnbull C, Al Olama AA, Dermitzakis ET, Linger R, Huddart RA, Renwick A, Hughes D, Hines S, Seal S, Morrison J, Nsengimana J, Deloukas P, Rahman N, Bishop DT, Easton DF, Stratton MR (2009) A genome-wide association study of testicular germ cell tumor. Nat Genet 41:807–810

Rescorla FJ, Breitfeld PP (1999) Pediatric germ cell tumors. Curr Probl Cancer 23:257–303

Richiardi L, Bellocco R, Adami HO, Torrang A, Barlow L, Hakulinen T, Rahu M, Stengrevics A, Storm H, Tretli S, Kurtinaitis J, Tyczynski JE, Akre O (2004) Testicular cancer incidence in eight northern European countries: secular and recent trends. Cancer Epidemiol Biomarkers Prev 13:2157–2166

Rodvall Y, Dich J, Wiklund K (2003) Cancer risk in offspring of male pesticide applicators in agriculture in Sweden. Occup Environ Med 60:798–801

Rosen A, Jayram G, Drazer M, Eggener SE (2011) Global trends in testicular cancer incidence and mortality. Eur Urol 60:374–379

Sarma A, McLaughlin J, Schottenfeld D (2006) Testicular cancer. In: Schottenfeld D, Fraumeni JE Jr (eds) Cancer epidemiology and prevention, 3rd edn. Oxford University Press, New York, pp 1159–1160

Schottenfeld D (1996) Testicular cancer. In: Schottenfeld D, Fraumeni JE Jr (eds) Cancer epidemiology and prevention, 2nd edn. Oxford University Press, New York, pp 1207–1219

Schottenfeld D, Warshauer ME, Sherlock S, Zauber AG, Leder M, Payne R (1980) The epidemiology of testicular cancer in young adults. Am J Epidemiol 112:232–246

Shah JP, Kumar S, Bryant CS, Ali-Fehmi R, Malone JM Jr, Deppe G, Morris RT (2008) A population-based analysis of 788 cases of yolk sac tumors: a comparison of males and females. Int J Cancer 123:2671–2675

Shah MN, Devesa SS, Zhu K, McGlynn KA (2007) Trends in testicular germ cell tumours by ethnic group in the United States. Int J Androl 30:206–213; discussion 213–214

Shankar S, Davies S, Giller R, Krailo M, Davis M, Gardner K, Cai H, Robison L, Shu XO (2006) In utero exposure to female hormones and germ cell tumors in children. Cancer 106:1169–1177

Shu XO, Nesbit ME, Buckley JD, Krailo MD, Robinson LL (1995) An exploratory analysis of risk factors for childhood malignant germ-cell tumors: report from the Childrens Cancer Group (Canada, United States). Cancer Causes Control 6:187–198

Shulman LP, Muram D, Marina N, Jones C, Portera JC, Wachtel SS, Simpson JL, Elias S (1994) Lack of heritability in ovarian germ cell malignancies. Am J Obstet Gynecol 170:1803–1805; discussion 1805–1808

Skakkebaek NE, Holm M, Hoei-Hansen C, Jorgensen N, Rajpert-De Meyts E (2003) Association between testicular dysgenesis syndrome (TDS) and testicular neoplasia: evidence from 20 adult patients with signs of maldevelopment of the testis. APMIS 111:1–9; discussion 9–11

Skakkebaek NE, Rajpert-De Meyts E, Main KM (2001) Testicular dysgenesis syndrome: an increasingly common developmental disorder with environmental aspects. Hum Reprod 16:972–978

Smith HO, Berwick M, Verschraegen CF, Wiggins C, Lansing L, Muller CY, Qualls CR (2006) Incidence and survival rates for female malignant germ cell tumors. Obstet Gynecol 107:1075–1085

Sonneveld DJ, Sleijfer DT, Schrafford Koops H, Sijmons RH, van der Graaf WT, Sluiter WJ, Hoekstra HJ (1999) Familial testicular cancer in a single-centre population. Eur J Cancer 35:1368–1373

Spermon JR, Witjes JA, Nap M, Kiemeney LA (2001) Cancer incidence in relatives of patients with testicular cancer in the eastern part of The Netherlands. Urology 57:747–752

Starr JR, Chen C, Doody DR, Hsu L, Ricks S, Weiss NS, Schwartz SM (2005) Risk of testicular germ cell cancer in relation to variation in maternal and offspring cytochrome p450 genes involved in catechol estrogen metabolism. Cancer Epidemiol Biomarkers Prev 14:2183–2190

Stephansson O, Wahnstrom C, Pettersson A, Sorensen HT, Tretli S, Gissler M, Troisi R, Akre O, Grotmol T (2011) Perinatal risk factors for childhood testicular germ-cell cancer: a Nordic population-based study. Cancer Epidemiol 35:e100–e104

Sun M, Abdollah F, Liberman D, Abdo A, Thuret R, Tian Z, Shariat SF, Montorsi F, Perrotte P, Karakiewicz PI (2011) Racial disparities and socioeconomic status in men diagnosed with testicular germ cell tumors: a survival analysis. Cancer 117:4277–4285

Swerdlow AJ, Huttly SR, Smith PG (1987) Prenatal and familial associations of testicular cancer. Br J Cancer 55:571–577

Swerdlow AJ, Stiller CA, Wilson LM (1982) Prenatal factors in the aetiology of testicular cancer: an epidemiological study of childhood testicular cancer deaths in Great Britain, 1953–73. J Epidemiol Community Health 36:96–101

Toriola AT, Surcel HM, Lundin E, Schock H, Grankvist K, Pukkala E, Chen T, Toniolo P, Lehtinen M, Zeleniuch-Jacquotte A, Lukanova A (2011) Insulin-like growth factor-I and C-reactive protein during pregnancy and maternal risk of non-epithelial ovarian cancer: a nested case–control study. Cancer Causes Control 22:1607–1611

Turnbull C, Rapley EA, Seal S, Pernet D, Renwick A, Hughes D, Ricketts M, Linger R, Nsengimana J, Deloukas P, Huddart RA, Bishop DT, Easton DF, Stratton MR, Rahman N (2010) Variants near DMRT1, TERT and ATF7IP are associated with testicular germ cell cancer. Nat Genet 42:604–607

United Kingdom Testicular Cancer Study Group (1994) Aetiology of testicular cancer: association with congenital abnormalities, age at puberty, infertility, and exercise. BMJ 308:1393–1399

Visvader JE, Lindeman GJ (2008) Cancer stem cells in solid tumours: accumulating evidence and unresolved questions. Nat Rev Cancer 8:755–768

Von Behren J, Spector LG, Mueller BA, Carozza SE, Chow EJ, Fox EE, Horel S, Johnson KJ, McLaughlin C, Puumala SE, Ross JA, Reynolds P (2011) Birth order and risk of childhood cancer: a pooled analysis from five US States. Int J Cancer 128:2709–2716

Walker AH, Ross RK, Haile RW, Henderson BE (1988) Hormonal factors and risk of ovarian germ cell cancer in young women. Br J Cancer 57:418–422

Walsh TJ, Dall'Era MA, Croughan MS, Carroll PR, Turek PJ (2007) Prepubertal orchiopexy for cryptorchidism may be associated with lower risk of testicular cancer. J Urol 178:1440–1446; discussion 1446

Walsh TJ, Davies BJ, Croughan MS, Carroll PR, Turek PJ (2008) Racial differences among boys with testicular germ cell tumors in the United States. J Urol 179: 1961–1965

Walsh TJ, Grady RW, Porter MP, Lin DW, Weiss NS (2006) Incidence of testicular germ cell cancers in U.S. children: SEER program experience 1973 to 2000. Urology 68:402–405; discussion 405

Wanderas EH, Grotmol T, Fossa SD, Tretli S (1998) Maternal health and pre- and perinatal characteristics in the etiology of testicular cancer: a prospective population- and register-based study on Norwegian males born between 1967 and 1995. Cancer Causes Control 9:475–486

Weir HK, Marrett LD, Kreiger N, Darlington GA, Sugar L (2000) Pre-natal and peri-natal exposures and risk of testicular germ-cell cancer. Int J Cancer 87: 438–443

Weir HK, Marrett LD, Moravan V (1999) Trends in the incidence of testicular germ cell cancer in Ontario by histologic subgroup, 1964–1996. Cmaj 160: 201–205

Weiss NS, Cook LS, Farrow DC, Rosenblatt KA (1996) Ovarian cancer. In: Schottenfeld D, Fraumeni JE Jr (eds) Cancer epidemiology and prevention, 2nd edn. Oxford University Press, New York, pp 1040–1057

Westergaard T, Olsen JH, Frisch M, Kroman N, Nielsen JW, Melbye M (1996) Cancer risk in fathers and brothers of testicular cancer patients in Denmark. A population-based study. Int J Cancer 66:627–631

Westhoff C, Pike M, Vessey M (1988) Benign ovarian teratomas: a population-based case–control study. Br J Cancer 58:93–98

Wilhelm D, Palmer S, Koopman P (2007) Sex determination and gonadal development in mammals. Physiol Rev 87:1–28

Wilkinson JD, Gonzalez A, Wohler-Torres B, Fleming LE, MacKinnon J, Trapido E, Button J, Peace S (2005) Cancer incidence among Hispanic children in the United States. Rev Panam Salud Publica 18:5–13

Wong TT, Ho DM, Chang KP, Yen SH, Guo WY, Chang FC, Liang ML, Pan HC, Chung WY (2005) Primary pediatric brain tumors: statistics of Taipei VGH, Taiwan (1975–2004). Cancer 104:2156–2167

Pathology of Germ Cell Tumors

Dinesh Rakheja and Lisa A. Teot

Contents

D. Rakheja (✉)
Department of Pathology,
The University of Texas Southwestern
Medical Center, Dallas, TX, USA
e-mail: dinesh.rakheja@utsouthwestern.edu

L.A. Teot (✉)
Department of Pathology,
Dana-Farber Cancer Institute,
Boston, MA, USA
e-mail: lisa.teot@childrens.harvard.edu

3.1   Yolk Sac Tumor

In male infants and toddlers, yolk sac tumor is the most commonly occurring pure germ cell tumor of the testis. In children, it is also the most commonly occurring malignant germ cell tumor component in an otherwise immature teratoma.

3.1.1   Gross Appearance

Grossly, yolk sac tumors are well-circumscribed, nonencapsulated, soft to firm, homogeneous tumors with a gelatinous appearance. Areas of hemorrhage, necrosis, or small cyst-like spaces may be seen (Fig. 3.1).

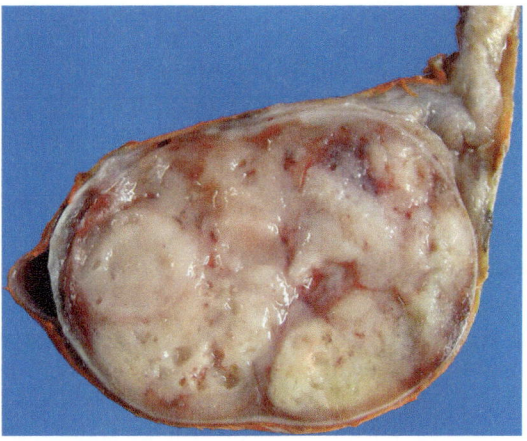

Fig. 3.1 Yolk sac tumor involving the testis. The tumor appears well circumscribed, lobulated, and pale tan with foci of hemorrhage and small cyst-like spaces

A.L. Frazier, J.F. Amatruda (eds.), *Pediatric Germ Cell Tumors*, Pediatric Oncology 1,
DOI 10.1007/978-3-642-38971-9_3, © Springer-Verlag Berlin Heidelberg 2014

Fig. 3.2 Yolk sac tumor
with characteristic hyaline
globules

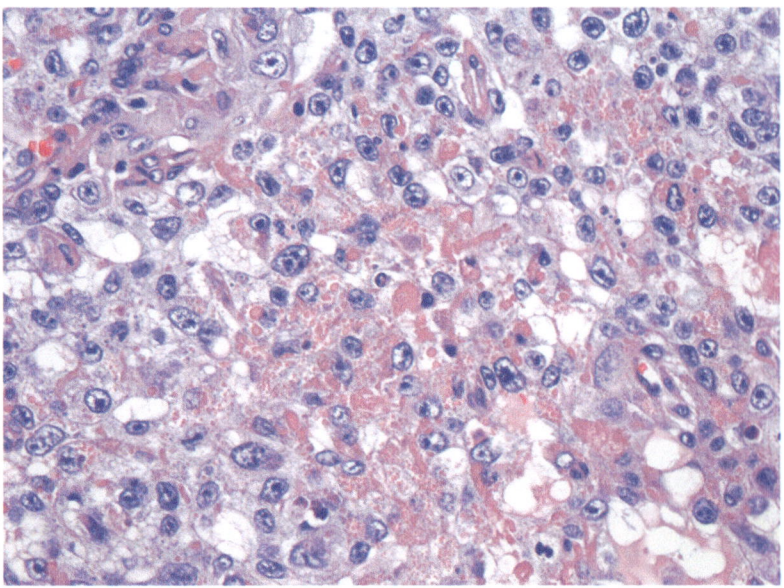

3.1.2   Microscopic Appearance

Yolk sac tumors are remarkably heterogeneous in microscopic appearance both within a tumor and between different tumors. They show architectural features that suggest differentiation toward extraembryonal endoderm (yolk sac, allantois) or somatic endoderm (lung, intestine, liver). Cytologically, the tumor cells are relatively uniform with round, polygonal, cuboidal/columnar, or oval shape; high nuclear-to-cytoplasmic ratio; clear, vacuolated, or pale eosinophilic cytoplasm; mild nuclear pleomorphism with variably prominent nucleoli; indistinct cell borders; and variable mitotic activity. The most common architecture is microcytic or reticular, where the cells are arranged in sheets, cords, or tubules; contain intracytoplasmic vacuoles of variable size; and have eccentrically placed nuclei. Characteristic intracytoplasmic and/or extracytoplasmic hyaline globules may be seen; the globules are highlighted by periodic acid-Schiff stain and are positive for alpha-fetoprotein (AFP) by immunohistochemistry (Fig. 3.2). The coalescence of microcysts may lead to large cystic spaces (macrocystic pattern, Fig. 3.3). Although

not always present, another characteristic feature of yolk sac tumors is Schiller-Duval bodies. These are glomeruloid structures with a central blood vessel surrounded by cuboidal tumor cells, which are surrounded by a cystic space and then a peripheral rim of flattened tumor cells (Fig. 3.4). Endodermal sinus pattern of yolk sac tumor is composed of many Schiller-Duval bodies. Solid pattern of yolk sac tumor is composed of sheets of tumor cells with pale eosinophilic or clear cytoplasm and without cystic spaces (Fig. 3.5). When the tumor cells are arranged in sheets and contain abundant eosinophilic cytoplasm, they may resemble primitive liver cells (hepatoid pattern). A few other architectural patterns include papillary/tubulopapillary (Fig. 3.6), micropapillary (Fig. 3.7), and glandular/alveolar (Fig. 3.8). Enteric or endometrioid pattern is a type of glandular pattern where the columnar cells have abundant clear cytoplasm and often subnuclear vacuoles (Fig. 3.9). Polyvesicular vitelline pattern of yolk sac tumor is composed of cystic or vesicular structures that show focal constriction and are lined by flattened to cuboidal tumor cells; the vesicles are haphazardly distributed in abundant myxoid to fibrous stroma

Fig. 3.3 Yolk sac tumor, macrocystic pattern

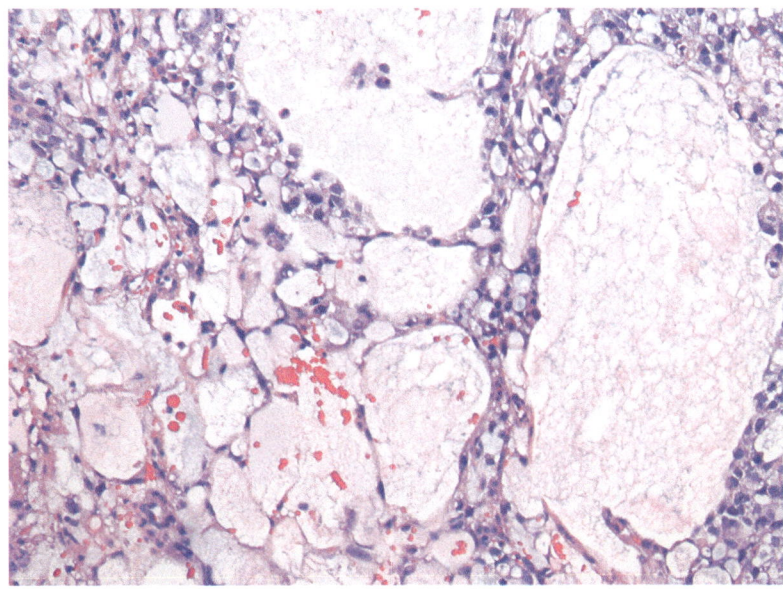

Fig. 3.4 Yolk sac tumor with characteristic Schiller-Duval bodies

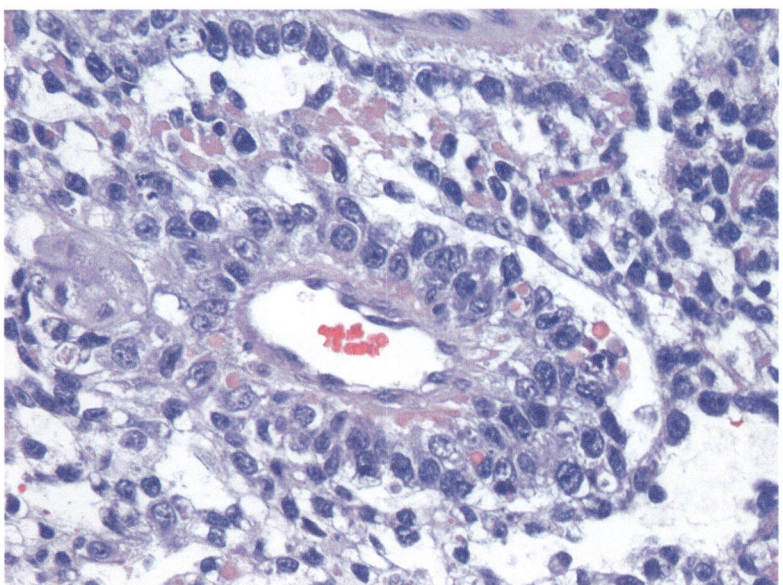

(Fig. 3.10). Myxomatous pattern is composed of abundant pale basophilic myxoid stroma with only scattered tumor cells (Fig. 3.11). Parietal pattern is composed of densely eosinophilic bands of basement membrane-like material dissecting in between tumor cells (Fig. 3.12).

3.1.3 Immunohistochemistry

Yolk sac tumors are positive for Lin28, SALL4, glypican-3, AFP, and cytokeratin and are occasionally focally positive for placental alkaline

Fig. 3.5 Yolk sac tumor, solid pattern

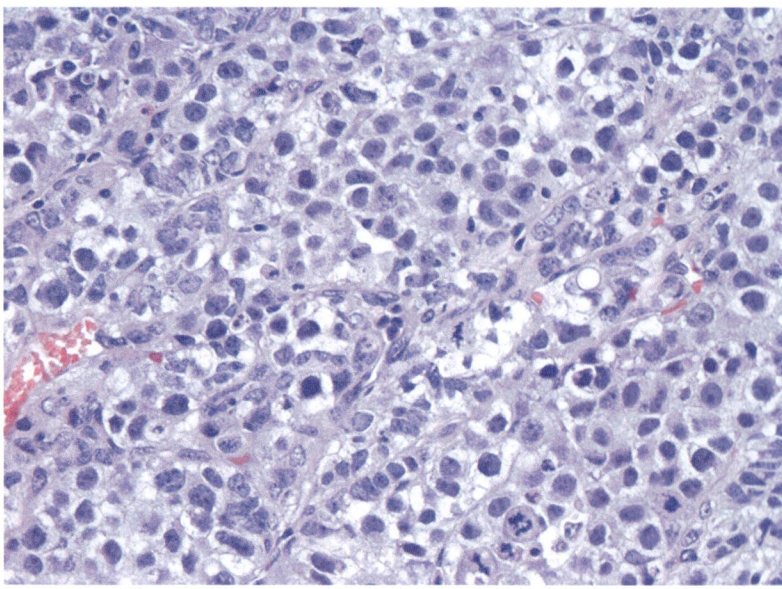

Fig. 3.6 Yolk sac tumor, papillary/tubulopapillary pattern

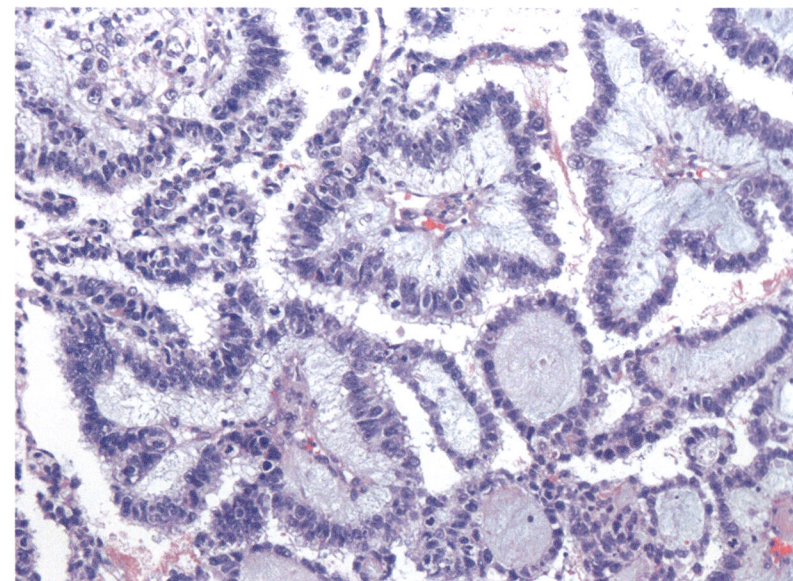

phosphatase (PLAP) (Fig. 3.13). They are usually negative for OCT3/4, c-kit (CD117), podoplanin (D2-40), CD30, and human chorionic gonadotropin (HCG).

3.2 Seminoma

In adults, seminoma is the most commonly occurring pure germ cell tumor. While it most commonly

Fig. 3.7 Yolk sac tumor, micropapillary pattern

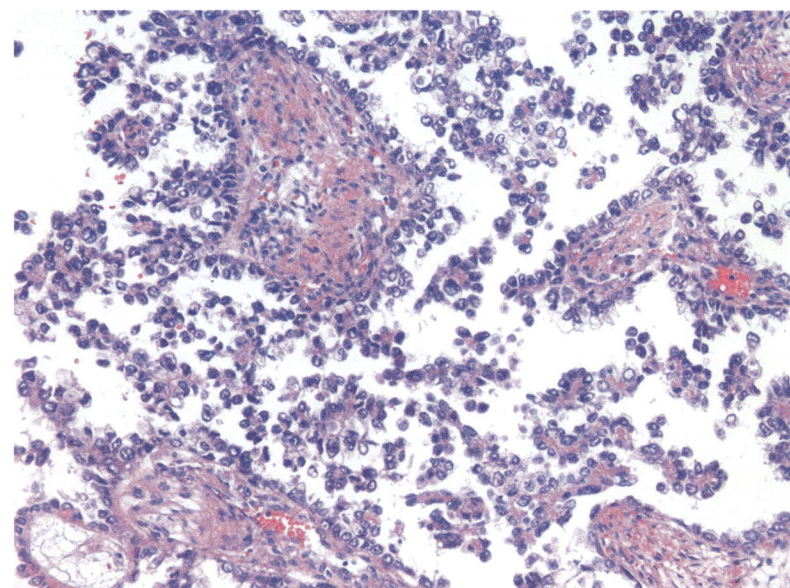

Fig. 3.8 Yolk sac tumor, glandular/alveolar pattern

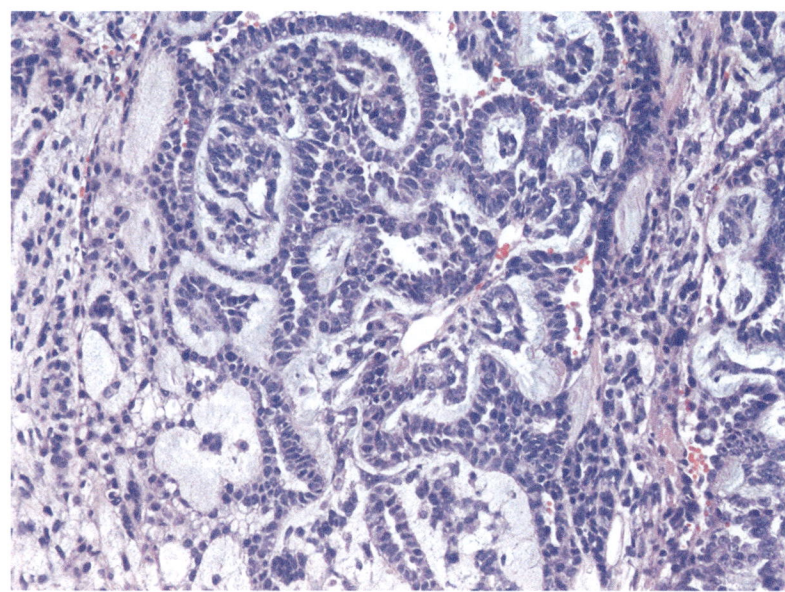

occurs in fourth and fifth decade, it does occur in older children. The term seminoma is used for the tumors occurring in testes, while identical tumors occurring in ovaries or dysgenetic gonads are termed dysgerminoma and those occurring in extragonadal sites are called germinomas.

3.2.1 Gross Appearance

Grossly, seminomas are well-circumscribed, nonencapsulated, lobulated, soft to firm, fleshy-appearing tumors that are typically uniformly tan to gray white (Fig. 3.14). Areas of necrosis and

Fig. 3.9 Yolk sac tumor, enteric or endometrioid pattern with subnuclear vacuoles

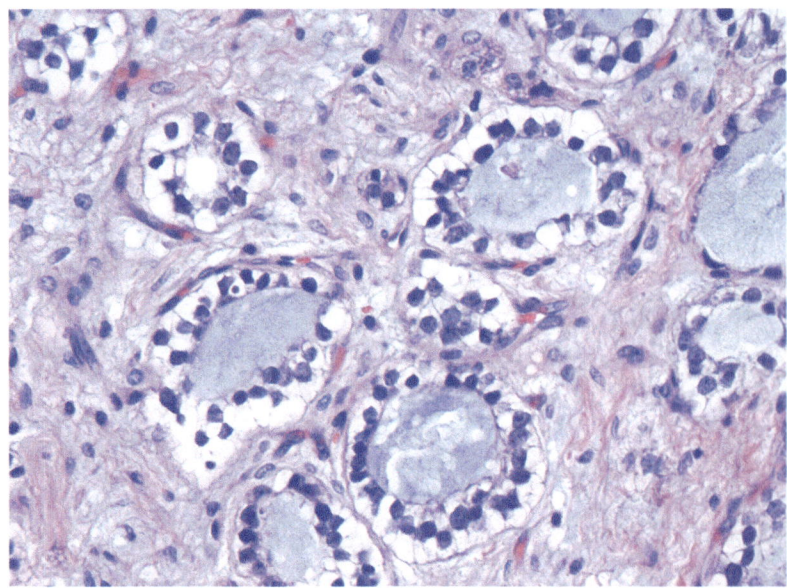

Fig. 3.10 Yolk sac tumor, polyvesicular vitelline pattern

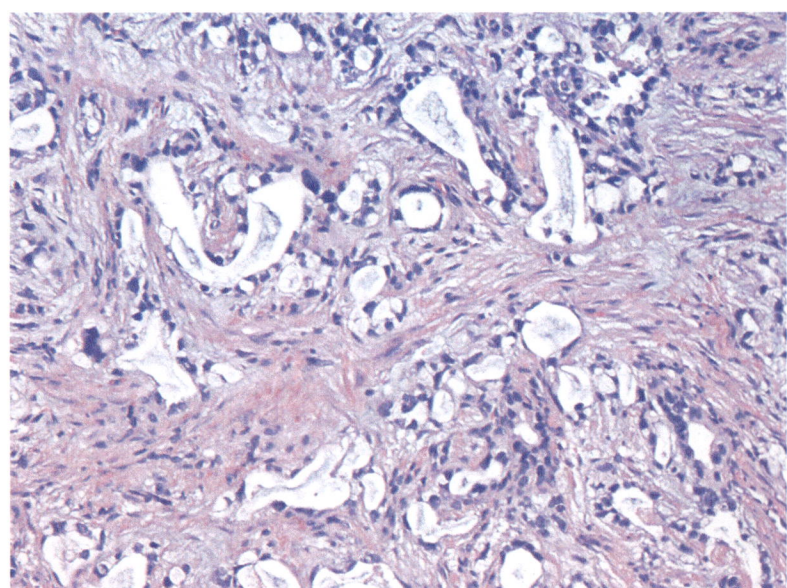

hemorrhage may indicate the presence of a non-seminomatous germ cell tumor component, but pure seminomas, particularly large tumors, may also show hemorrhage and necrosis.

3.2.2 Microscopic Appearance

The typical seminomas are composed of tumor cells arranged in monotonous sheets that are divided into variably sized nests by fibrovascular septa containing variable amounts of lympho-plasmacytic inflammation. The tumor cells appear primitive and morphologically resemble primordial germ cells or gonocytes. They are usually uniform, large, round to polygonal, with well-defined cell membranes, abundant clear cytoplasm, round central nucleus, 1–2 prominent but small nucleoli, and variable mitotic activity (Fig. 3.15). Occasional tumors show increased

Fig. 3.11 Yolk sac tumor, myxomatous pattern with few tumor cells in abundant myxoid stroma

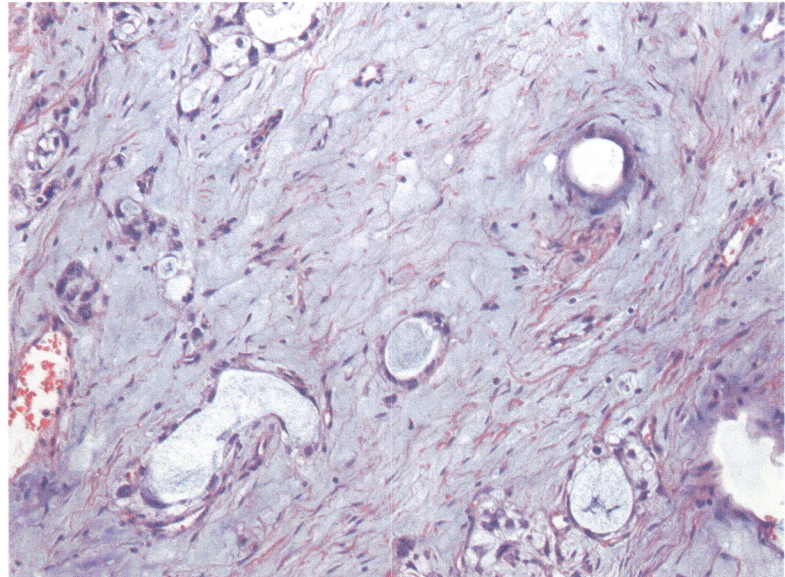

Fig. 3.12 Yolk sac tumor, parietal pattern with tumor cells separated by bands of dense, basement membrane-like material

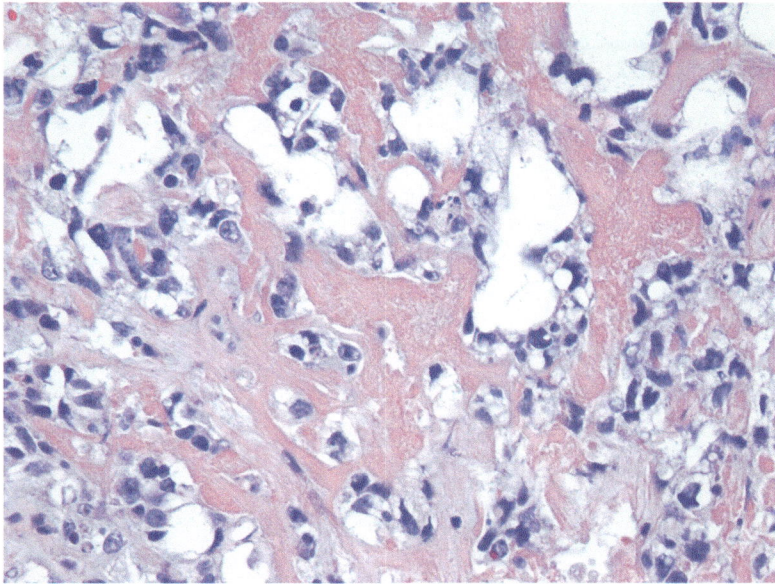

pleomorphism, anisonucleosis, increased nuclear-to-cytoplasmic ratio, larger nucleoli, and increased mitoses. These so-called anaplastic seminomas have no apparent prognostic impact on the disease course (Fig. 3.16). Rare seminomas show trabecular architecture (Fig. 3.17), pseudoglandular architecture (Fig. 3.18), or foci of cystic degeneration. The fibrovascular septa may show lymphoid follicles (Fig. 3.19) or granulomatous inflammation (Fig. 3.20). Rarely, the fibrous septa may coalesce and the fibrosis may become diffuse, giving rise to a sclerotic appearance (Fig. 3.21). Dysgerminomas may overgrow gonadoblastomas in dysgenetic gonads (Fig. 3.22). Intratubular germ cell neoplasia (ITGCN, seminoma-like cells distributed along the periphery of seminiferous tubules) and intratubular seminomas (seminoma-like cells filling and expanding the seminiferous tubules) may be seen adjacent to seminomas.

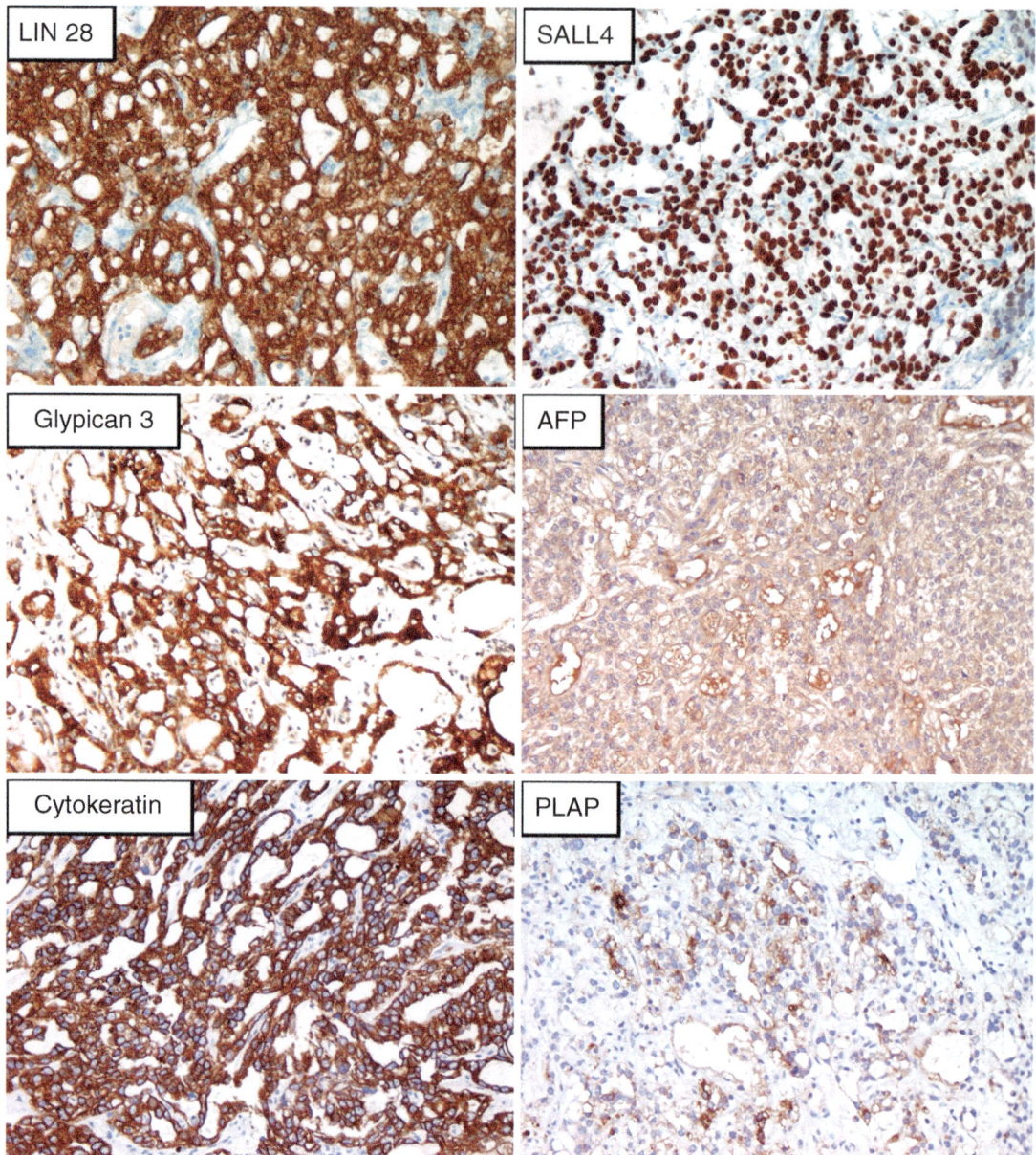

Fig. 3.13 Yolk sac tumor is immunoreactive for (**a**) Lin28, (**b**) SALL4, (**c**) glypican-3, (**d**) AFP, (**e**) cytokeratin, and (**f**) PLAP (focal)

3.2.3 Immunohistochemistry

Seminomas are positive for PLAP, c-kit (CD117), podoplanin (D2-40), OCT3/4, SALL4, and Lin28 (Fig. 3.23). They may show focal/weak reactivity for cytokeratin and CD30 and do not stain for glypican-3, AFP, or HCG. Rarely, HCG-positive syncytiotrophoblasts may be present in otherwise typical seminomas; scattered syncytiotrophoblasts by themselves are not indicative of a component of choriocarcinoma and do not affect the prognosis, but may lead to elevated levels of serum HCG. ITGCN and intratubular seminomas show staining patterns similar to seminomas.

3.3   Embryonal Carcinoma

In adults, embryonal carcinoma is the second most common pure germ cell neoplasm of the testis. In contrast, pure embryonal carcinoma of either the testis or ovary is exceedingly rare in children and adolescents, although embryonal carcinoma is a common component of mixed germ cell tumors.

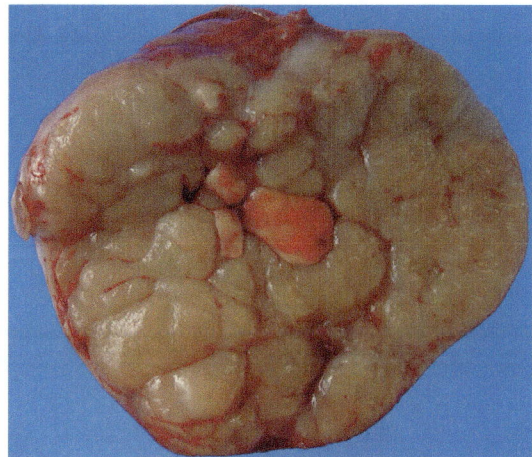

Fig. 3.14 Seminoma appears as a well-circumscribed, lobulated, fleshy, uniformly tan tumor

3.3.1   Gross Appearance

Grossly, embryonal carcinomas are nonencapsulated, often with ill-defined borders, soft to firm, inhomogeneous tumors with smooth to granular, gray white cut surfaces that bulge above the adjacent testicular parenchyma. Areas of hemorrhage and necrosis are often present (Fig. 3.24).

3.3.2   Microscopic Appearance

Embryonal carcinomas are composed of malignant cells arranged in solid, glandular, and papillary patterns with foci of hemorrhage, necrosis, and fibrosis (Fig. 3.25). The malignant cells are pleomorphic, large, and epithelioid, with poorly defined cell membranes, abundant amphophilic cytoplasm, and large, irregular nuclei with vesicular chromatin and large, prominent nucleoli (Fig. 3.26). Particularly in areas with solid architecture, indistinct cytoplasmic borders and nuclear overlapping impart a syncytial appearance. Mitotic activity is brisk and abnormal mitotic figures are usually present. Giant anaplastic tumor cells may be present. In some tumors, fibrous stroma is prominent and may vary from

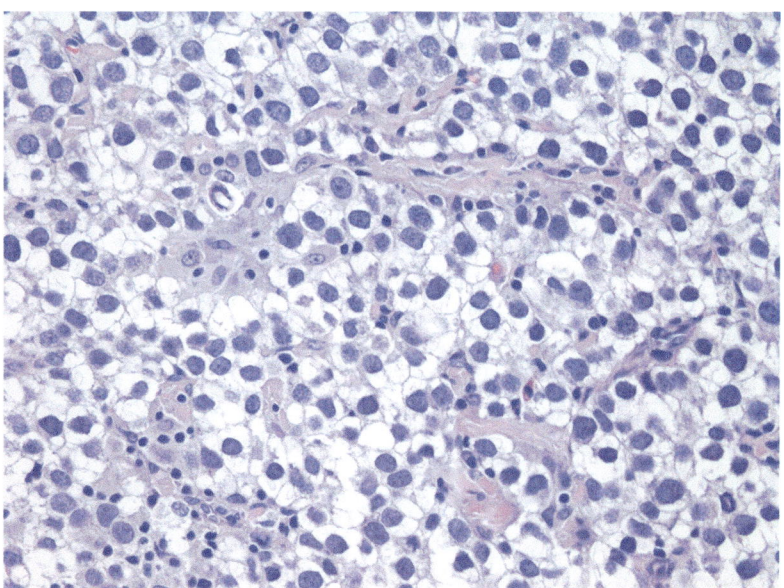

Fig. 3.15 Seminoma is composed of uniform, large, round to polygonal cells, with well-defined cytoplasmic membranes, abundant clear cytoplasm, round central nuclei, and 1–2 prominent but small nucleoli

Fig. 3.16 Seminoma with anaplastic features including nuclear pleomorphism, anisonucleosis, increased nuclear-to-cytoplasmic ratio, larger nucleoli, and increased mitoses

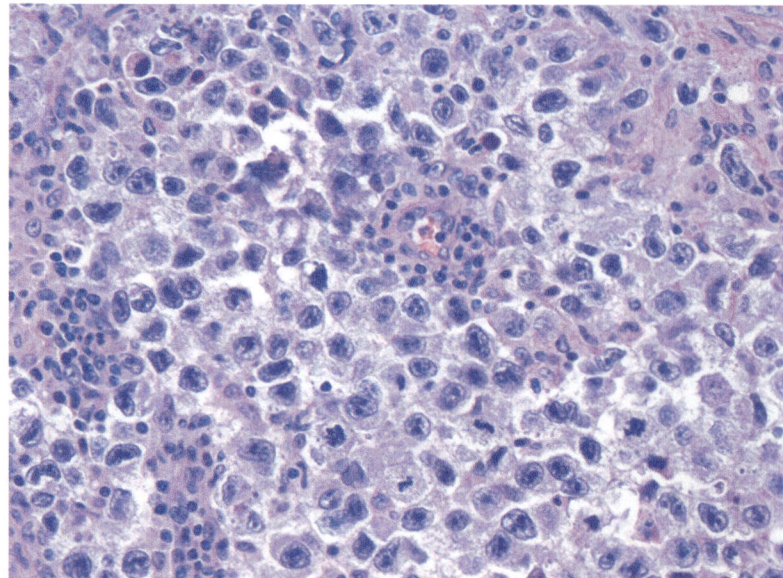

Fig. 3.17 Seminoma with trabecular architecture

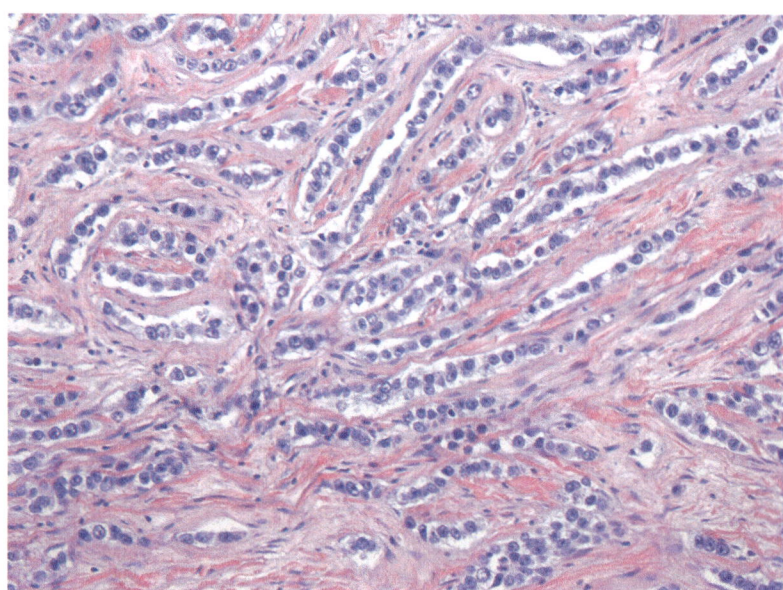

paucicellular to densely cellular (Fig. 3.27); however, the presence of cellular mesenchymal tissue by itself does not warrant a diagnosis of teratoma (Young 2008). Intratubular embryonal carcinoma (embryonal carcinoma filling and expanding the seminiferous tubules) is often seen within and adjacent to the tumor (Fig. 3.28).

3.3.3 Immunohistochemistry

Embryonal carcinomas are positive for CD30, PLAP, OCT3/4, SALL4, Lin28, and pancytokeratin (Fig. 3.29). They may show focal/weak reactivity for AFP and podoplanin (D2-40) and do not stain for EMA, glypican-3, c-kit (CD117), or

Fig. 3.18 Seminoma with pseudoglandular architecture

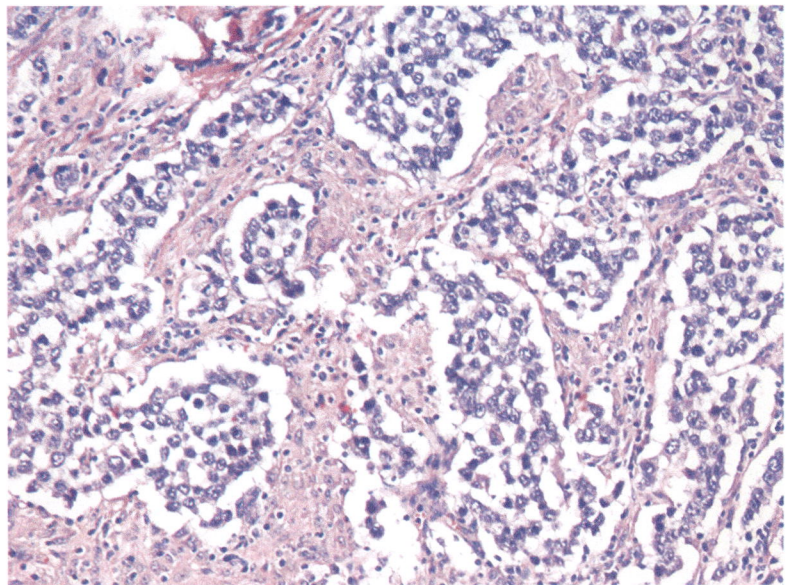

Fig. 3.19 Seminoma with lymphoid follicle

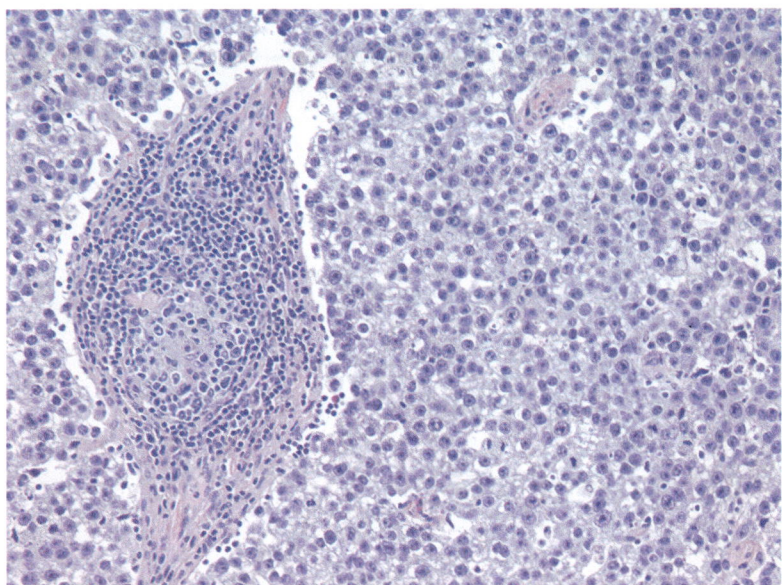

HCG. Rarely, HCG-positive syncytiotrophoblasts may be present in otherwise typical embryonal carcinomas; scattered syncytiotrophoblasts in the absence of cytotrophoblasts are not indicative of a component of choriocarcinoma and do not affect the prognosis, but may lead to elevated levels of serum HCG.

3.4 Mature and Immature Teratoma

Teratomas arise in both gonadal and extragonadal sites, including the sacrococcygeal region, retroperitoneum, mediastinum, neck, and central nervous system. They may be pure or a component

Fig. 3.20 Seminoma with granulomatous inflammation

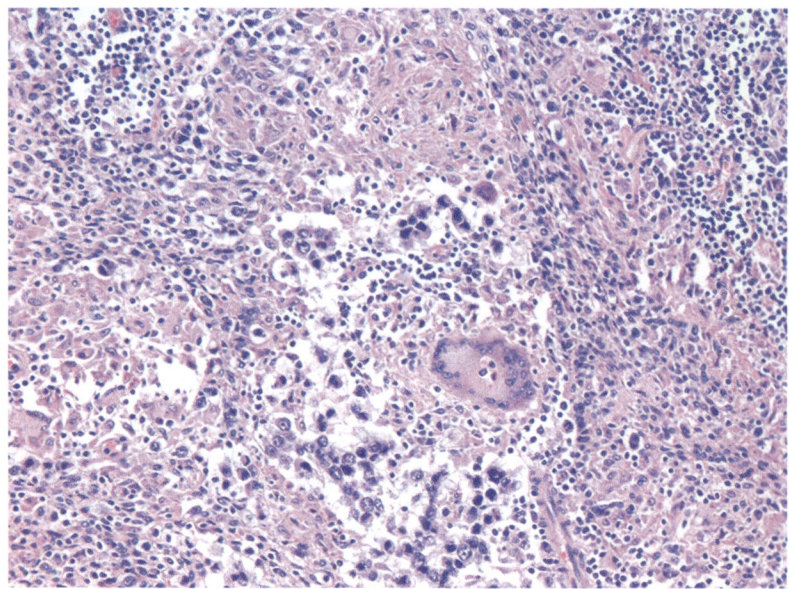

Fig. 3.21 Seminoma with diffuse fibrosis

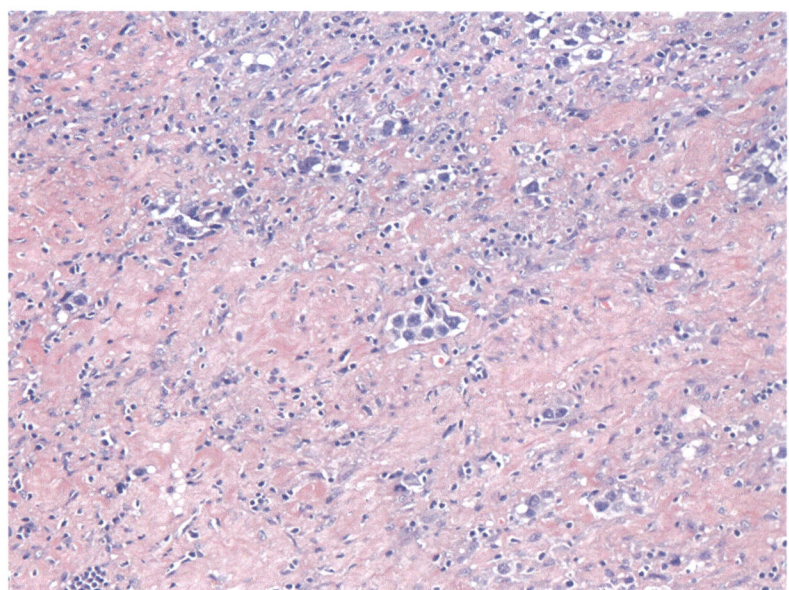

of mixed germ cell tumors and mature or immature. Behavior varies with the location of the tumor and age at presentation and, depending on these factors, may be independent of histology.

3.4.1 Gross Appearance

Teratomas may be predominantly cystic, solid and cystic, or predominantly solid tumors and are often very large. Tumors usually appear well demarcated from the surrounding uninvolved tissue, although the borders of some extragonadal tumors are less well defined. The cystic components may contain hair, sebaceous or keratinous material, or clear, serous, or mucinous fluid (Fig. 3.30). The solid components may be smooth, nodular or lobulated, gelatinous, soft, or firm, and white, gray, tan, red, or dark brown (Fig. 3.31). Tissues, such as cartilage, bone, teeth, brain, cerebellum, and pigmented retina, may be recognizable grossly. Predominantly

Fig. 3.22 Dysgerminoma overgrowing gonadoblastoma in dysgenetic gonad

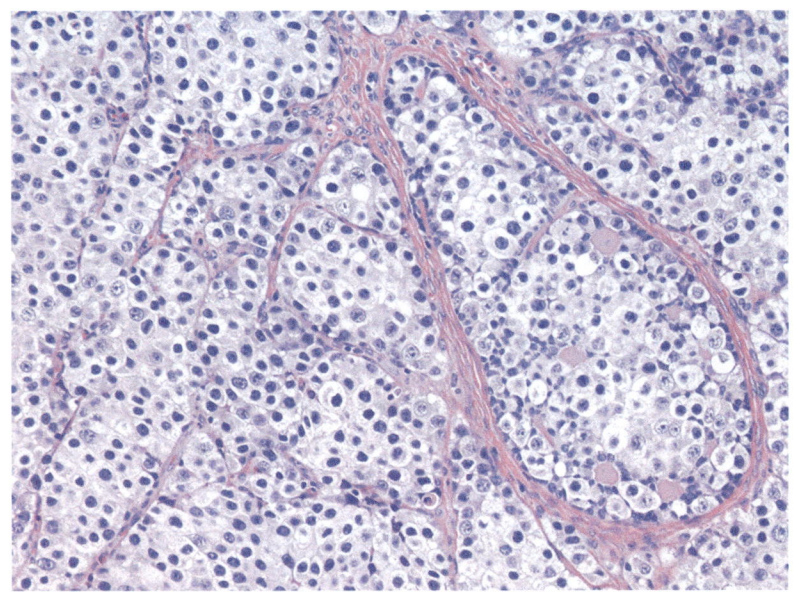

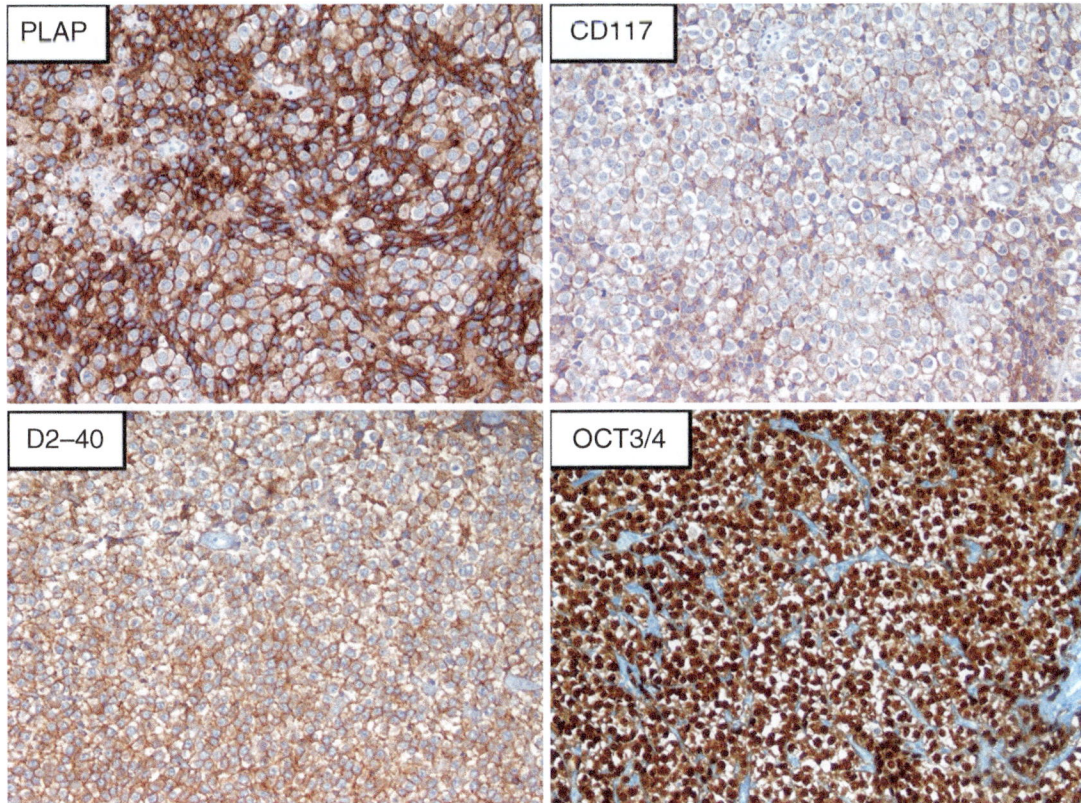

Fig. 3.23 Seminoma is immunoreactive for (**a**) PLAP, (**b**) c-kit (CD117), (**c**) podoplanin (D2-40), (**d**) OCT 3/4, (**e**) SALL4, and (**f**) Lin28

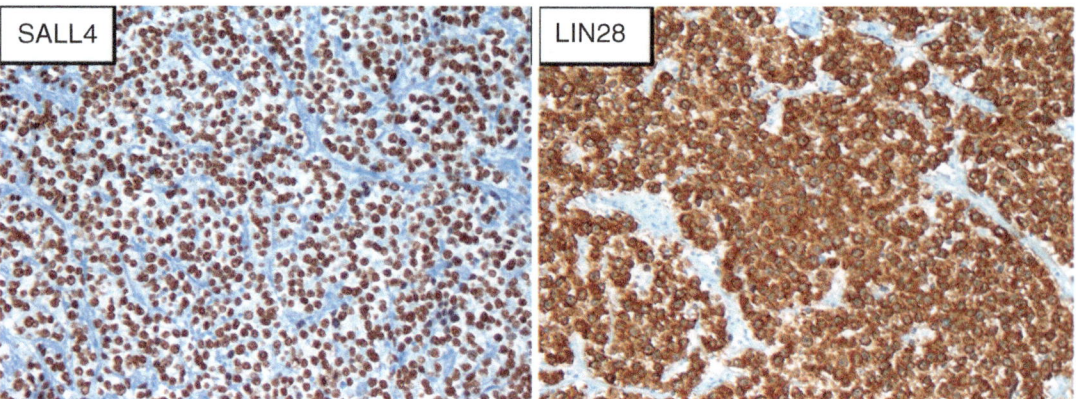

Fig. 3.23 (continued)

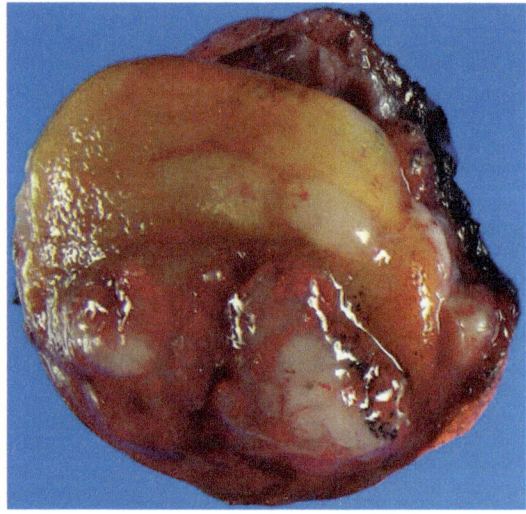

Fig. 3.24 Embryonal carcinoma involving the testis. The tumor appears nonencapsulated, smooth, gray white with focal hemorrhage and bulges above the adjacent testicular parenchyma

cystic teratomas are usually mature and infrequently contain immature or malignant elements. In contrast, solid and cystic and predominantly solid teratomas may be either mature or immature and should be extensively sampled to exclude the presence of malignant germ cell elements.

3.4.2 Microscopic Appearance

Teratomas are composed of haphazardly arrayed tissues representing the three germ layers, ectoderm, mesoderm, and endoderm. As the name implies, mature teratomas consist of mature elements that are easily recognized as histologically corresponding to their non-neoplastic counterparts (Fig. 3.32). Common ectodermal components include skin, adnexal structures, neuroglial tissue, choroid plexus, and teeth, while common endodermal elements include respiratory and digestive mucosa, pancreatic tissue, and thyroid. Mesodermal elements include adipose tissue, cartilage, bone, and skeletal and smooth muscle. The component tissues show varying degrees of organization and may recapitulate the structure of an organ. Immature teratomas are characterized by the presence of elements with histologic features of their fetal or embryonal counterparts. The most common immature elements are neuroectodermal tissue and cellular mesenchyme, but some tumors also contain immature epithelium, rhabdomyoblasts, or other primitive tissues. Immature neuroectodermal tissue consists of immature neuroepithelial cells arranged in sheets, columnar embryonal cells with stratified hyperchromatic nuclei arrayed in tubules recapitulating primitive neurotubules, neuroblasts, and immature glia (Fig. 3.33). Mitotic activity is present and often brisk. Histologic grading of immature teratomas is based on quantification of the immature neuroectodermal elements, with Grade 1 defined by rare foci occupying not more than one low-power field per slide, Grade 2 defined by multiple foci occupying up to three low-power fields per slide, and Grade 3 defined

Fig. 3.25 Embryonal carcinoma with areas of (**a**) solid and (**b**) glandular growth

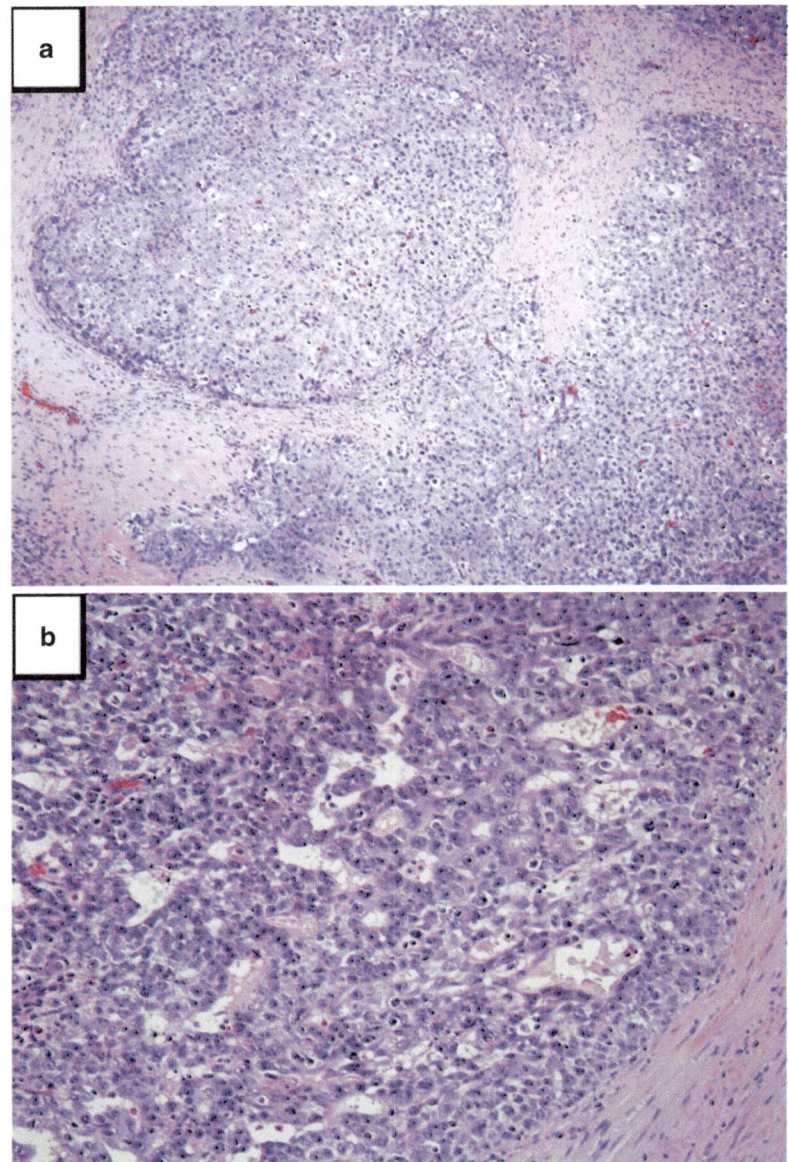

by numerous foci occupying four or more low-power fields per slide (Norris et al. 1976). Although this grading system is useful for immature ovarian teratomas in adults in whom prognosis is related to grade (O'Connor and Norris 1994), its applicability to immature teratomas of the ovary and other sites in the pediatric population is less clear, as prognosis appears to be independent of histologic grade (Heifetz et al. 1998; Marina et al. 1999; Cushing et al. 1999; Ross et al. 2002; Heerema-McKenney et al. 2005; Lu and Gershenson 2005). Immature teratomas may harbor malignant germ cell elements, most commonly yolk sac tumor, and, rarely, malignant somatic elements, such as rhabdomyosarcoma, primitive neuroectodermal tumor, nephroblastoma (Wilms tumor), or carcinoma.

3.4.3   Immunohistochemistry

The major role for immunohistochemistry in teratomas is to confirm the identity of suspected malignant elements (Fig. 3.34).

Fig. 3.26 Embryonal carcinoma is composed of pleomorphic, large, epithelioid cells with poorly defined cell membranes, abundant amphophilic cytoplasm, and large, irregular nuclei with vesicular chromatin and large, prominent nucleoli. Indistinct cytoplasmic borders and nuclear overlapping impart a syncytial appearance

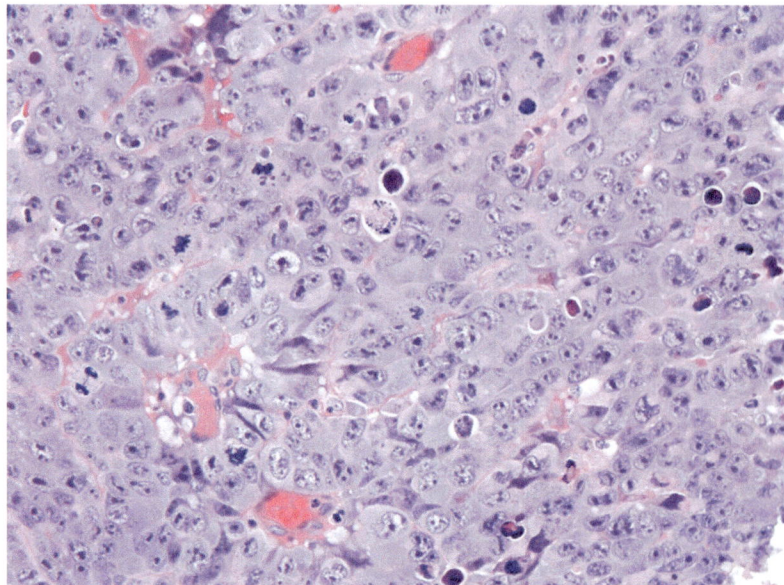

Fig. 3.27 Embryonal carcinoma with highly cellular stroma

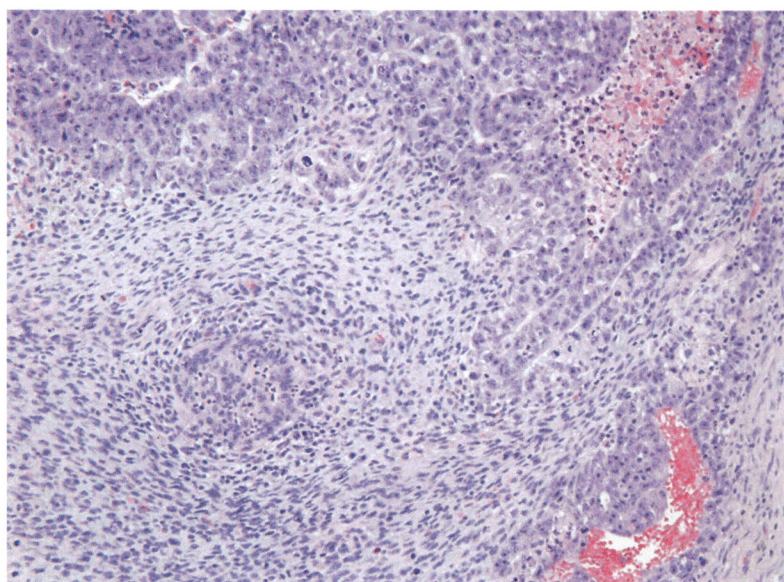

Fig. 3.28 Intratubular embryonal carcinoma fills and markedly expands a seminiferous tubule

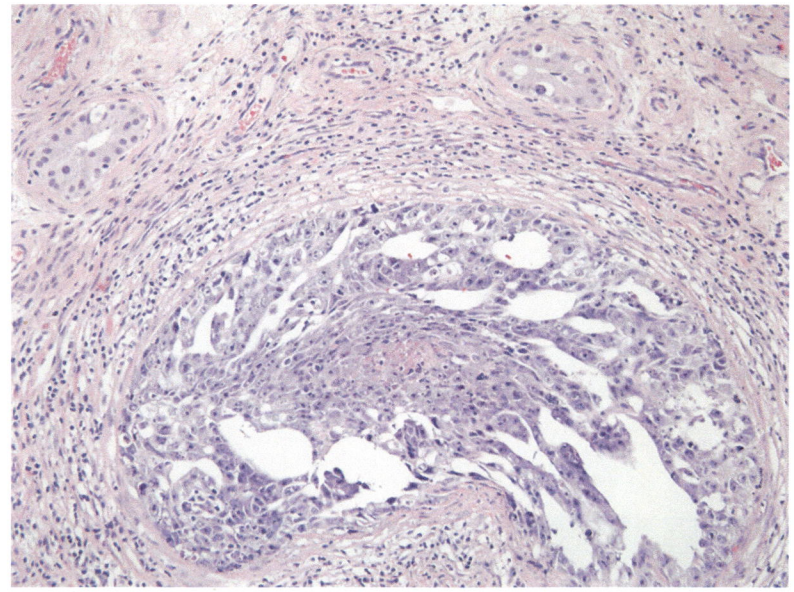

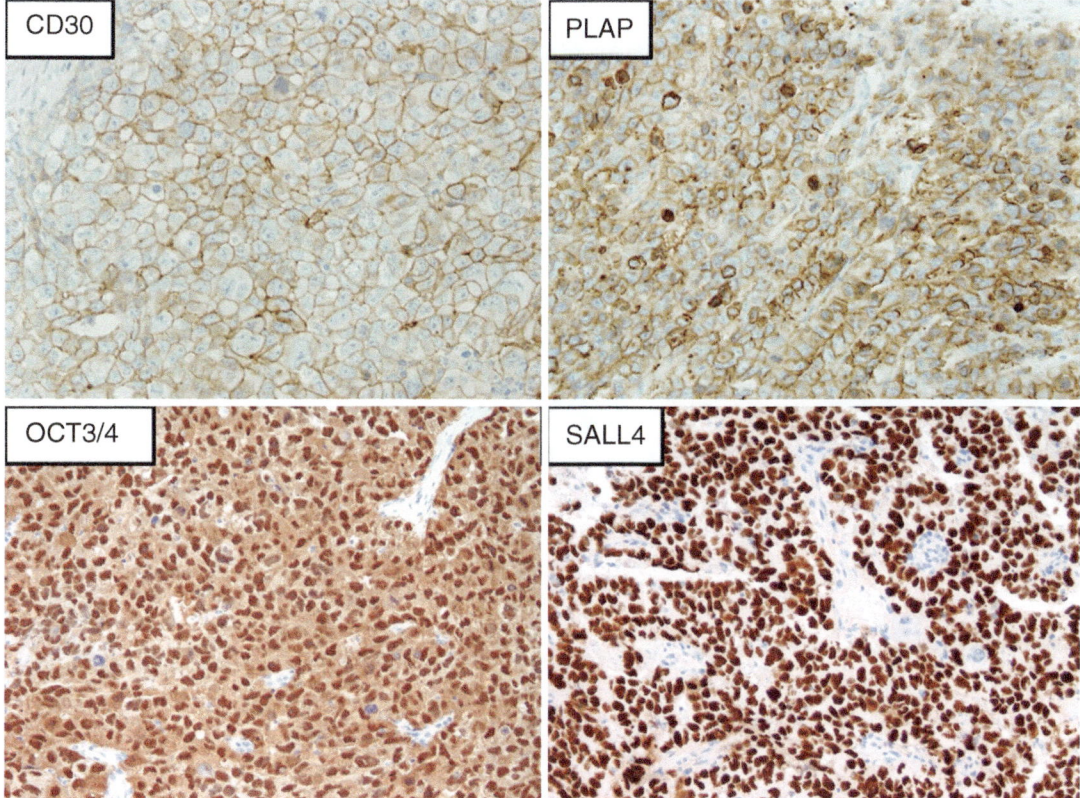

Fig. 3.29 Embryonal carcinoma is immunoreactive for (**a**) CD30, (**b**) PLAP, (**c**) OCT 3/4, (**d**) SALL4, (**e**) Lin28, and (**f**) cytokeratin

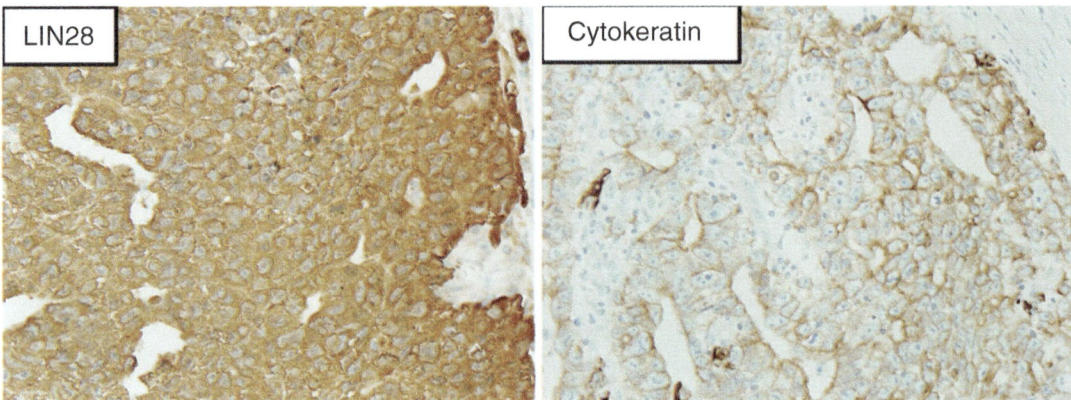

Fig. 3.29 (continued)

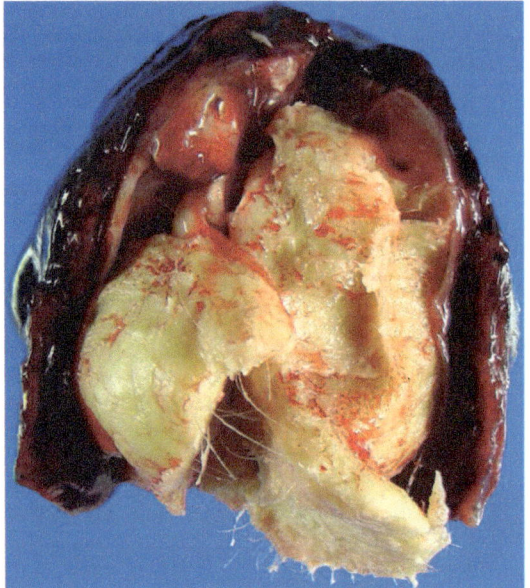

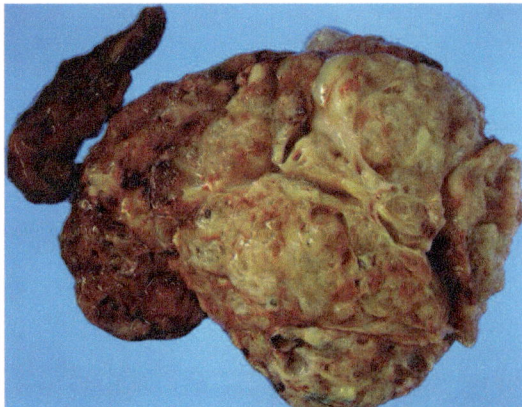

Fig. 3.31 Immature teratoma appearing as a predominantly solid tumor

Fig. 3.30 Mature teratoma with a prominent cystic component containing hair and sebaceous material

Fig. 3.32 Mature teratoma with (**a**) brain (neuroglial tissue), pancreatic tissue, and smooth muscle; (**b**) pancreatic tissue, cartilage, smooth muscle, pigmented retinal epithelium, and mucin-rich intestinal epithelium; and (**c**) brain, bone, and cystic spaces lined by nonkeratinizing squamous, ciliated respiratory, and cuboidal epithelium

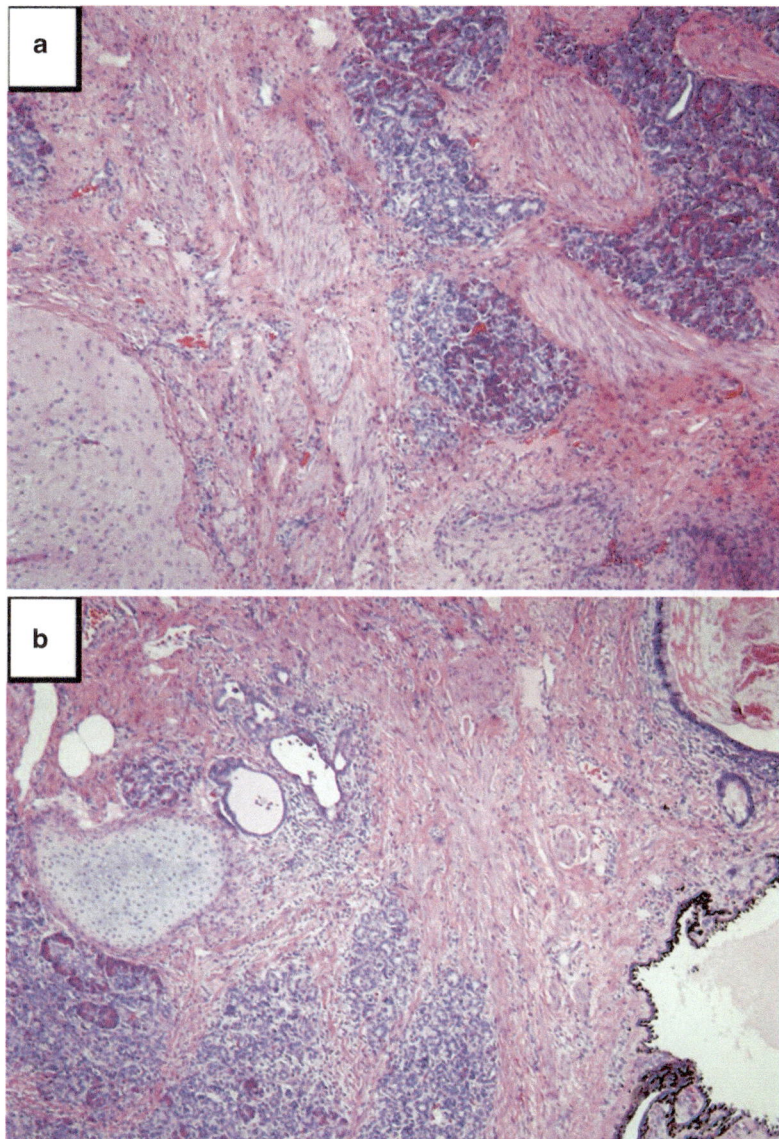

Fig. 3.32 (continued)

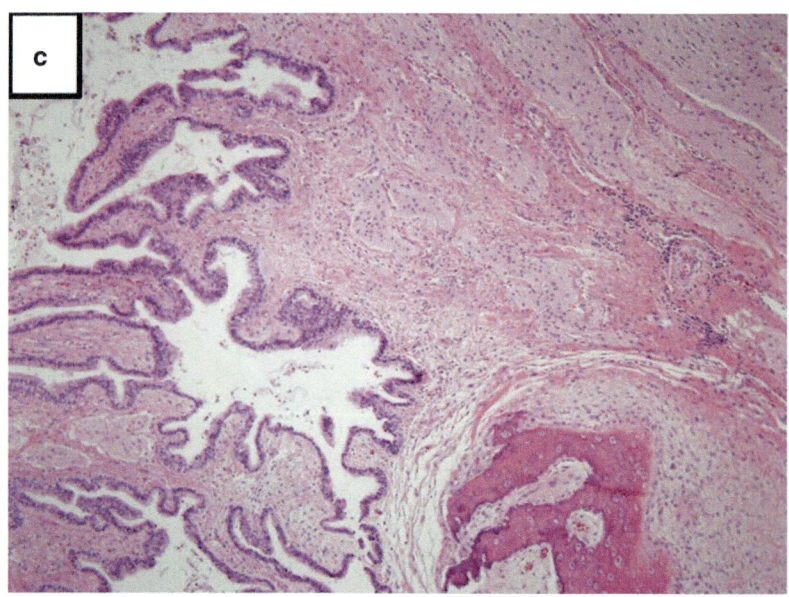

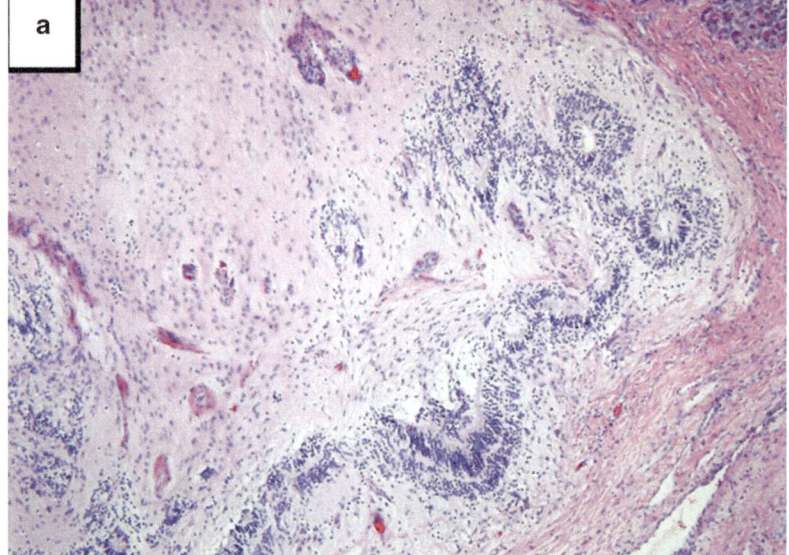

Fig. 3.33 Immature teratoma, (**a**) Grade 1, characterized by this small focus of immature neural tissue, and (**b**) Grade 3, with abundant, immature neural tissue forming primitive neural tubules

Fig. 3.33   (continued)

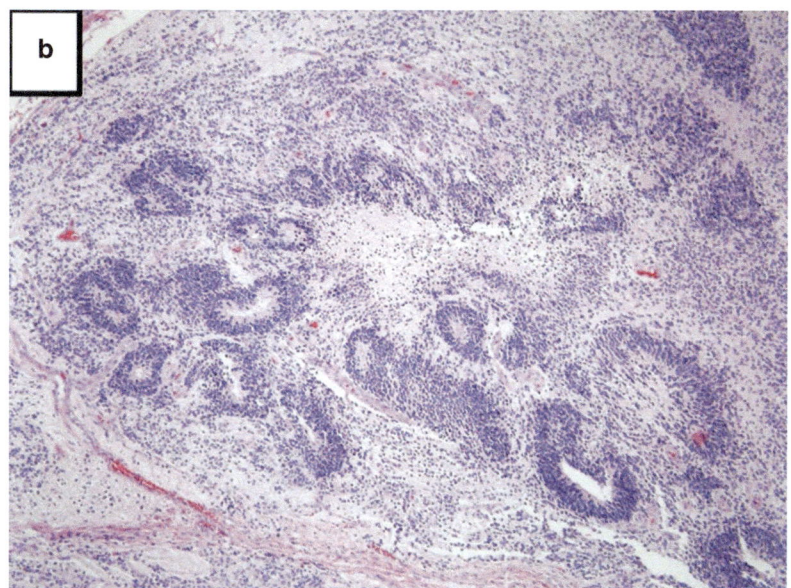

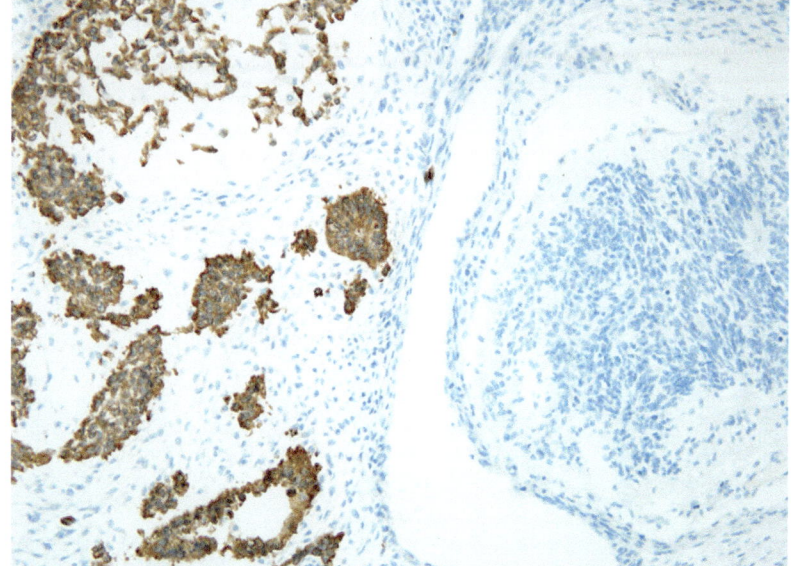

Fig. 3.34 An immunohistochemical stain for Lin28
highlights yolk sac tumor,
while the adjacent immature
neuroepithelial elements are
negative in this immature
teratoma with foci of yolk
sac tumor

3.5   Mixed Germ Cell Tumors

Mixed germ cell tumors are malignant neoplasms composed of two or more types of germ cell tumor and have histologic and immunophenotypic features characteristic of the various components, as discussed in Sects. 4.1, 4.2, 4.3, and 4.4.

References

Cushing B, Giller R, Ablin A et al (1999) Surgical resection alone is effective treatment for ovarian immature teratoma in children and adolescents: a report of the Pediatric Oncology Group and the Children's Cancer Group. Am J Obstet Gynecol 181:353–358

Heerema-McKenney A, Harrison MR, Bratton B et al (2005) Congenital teratoma: a clinicopathologic study of 22 fetal and neonatal tumors. Am J Surg Pathol 29:29–38

Heifetz SA, Cushing B, Giller R et al (1998) Immature teratomas in children: pathologic considerations: a report from the combined Pediatric Oncology Group/Children's Cancer Group. Am J Surg Pathol 22:1115–1124

Lu KH, Gershenson DM (2005) Update on the management of ovarian germ cell tumors. J Reprod Med 50:417–425

Marina NM, Cushing B, Giller R et al (1999) Complete surgical excision is effective treatment for children with immature teratomas with or without malignant elements: A Pediatric Oncology Group/Children's Cancer Group Intergroup Study. J Clin Oncol 17:2137–2143

Norris HL, Zirkin HJ, Benson WL (1976) Immature (malignant) teratoma of the ovary: a clinical and pathologic study of 58 cases. Cancer 37:2359–2372

O'Connor DM, Norris HJ (1994) The influence of grade on the outcome of stage I ovarian immature (malignant) teratomas and the reproducibility of grading. Int J Gynecol Pathol 13(4):283–289

Ross JH, Rybicki L, Kay R (2002) Clinical behavior and a contemporary management for prepubertal testis tumors: a summary of the Prepubertal testis Tumor Registry. J Urol 168:1675–1678; discussion 1678–1679

Young RH (2008) Testicular tumors – some new and a few perennial problems. Arch Pathol Lab Med 132:548–564

Clinical Treatment of Extracranial Pediatric Germ Cell Tumors

4

Furqan Shaikh and Juliet Hale

Contents

F. Shaikh (✉)
Division of Haematology/Oncology,
The Hospital for Sick Children,
Toronto, ON, Canada
e-mail: furqan.shaikh@sickkids.ca

J. Hale
Royal Victoria Infirmary, Newcastle upon Tyne, UK
e-mail: juliet.hale@nuth.nhs.uk

4.1 Introduction

Malignant germ cell tumors (MGCTs) account for 3.5 % of all pediatric cancers that occur before 15 years of age, making them approximately as common as childhood rhabdomyosarcomas, bone sarcomas, or retinoblastomas (Howlader et al. 2011). In children between 15 and 19 years of age, germ cell tumors account for 13.9 % of neoplasms, becoming the second most common malignancy (after Hodgkin's lymphoma) in this age group. Based on the most recent estimates from the Surveillance, Epidemiology, and End Results (SEER) Program of the National Cancer Institute, there are about 900 new cases of MGCT diagnosed in the United States each year among persons less than 20 years of age.

The dramatic increase in proportion with age is due to the sharp rise in the incidence of testicular germ cell tumors in postpubertal boys and young adult men (Wood et al. 2010). The natural consequence of this epidemiologic pattern is that a much larger sample size for clinical trials has been possible in adult testicular MGCTs, and thus much of what is first empirically discovered about risk stratification and treatment of MGCTs is based on investigations in this population.

The development of effective treatments for MGCTs has been one of the major successes of clinical oncology. In the 1950s–1970s, combination chemotherapy with various regimens such as vincristine, actinomycin D, and cyclophosphamide (VAC) produced occasional complete responses in MGCTs. However, the major breakthrough in the treatment of MGCT came in 1977, when Einhorn and Donohue added cisplatin to create a PVB regimen and obtained 100 % response and 64 % survival in men with disseminated testicular GCTs (Einhorn and Donohue 1977). This landmark report established the crucial role of cisplatin within combination chemotherapy regimens in treating germ cell tumors, and since then every successful regimen for treating MGCTs has contained a platinum compound. Etoposide was introduced in the 1980s, and early studies of the combination of cisplatin and etoposide for relapsed MGCTs represented the first time an adult solid tumor was cured with second-line chemotherapy (Einhorn 2002). This spawned the development of the PEB (cisplatin, etoposide, bleomycin) combination (Loehrer 1991), and a randomized study showed its superiority over the previous PVB regimen (Williams et al. 1987). Several subsequent randomized controlled trials further fine-tuned the risk stratification and duration of treatment, but PEB or PE remained established as the standard of care for adult testicular MGCTs.

Childhood MGCTs differ from the adult disease in many important ways, including the distribution of site, stage, histology, and genetics. Moreover, long-term toxicities of chemotherapy that may be acceptable in an adult patient can have more significant consequences for the young patient who is undergoing growth and development. Thus, while building on the results of studies in adult testicular MGCTs, pediatric oncology collaborative groups have tested, adapted, or modified these experiences in their own ways. An enduring challenge has been determining which of the findings from adult studies can be extrapolated to children and which require alternative approaches.

4.2 Development and Evolution of Treatment Regimens for Malignant Germ Cell Tumors in Children

Cure rates for the disease in children has improved dramatically over time, with overall survival (OS) rates increasing from about 40 % in the 1970s (Brodeur et al. 1981) to over 85 % in most recent trials. This success can be attributed to the development of a multidisciplinary treatment strategy utilizing both surgical tumor resection and systemic combination chemotherapy. Under select circumstances, one or the other modality alone can also be sufficient for cure.

MGCTs are chemosensitive tumors. Active single agents include platinum compounds (cisplatin, carboplatin, oxaliplatin), etoposide, vinblastine, bleomycin, ifosfamide, cyclophosphamide, actinomycin D, and paclitaxel. The commonly used combination regimens, and their corresponding acronyms, are listed in the abbreviations in Tables 4.1, 4.2, 4.3, 4.4, and 4.5.

This chapter will review the published experiences and outcomes of trials for childhood MGCTs across some of the pediatric oncology cooperative groups. It is important to mention at the outset that each group has historically used different systems for staging, risk stratification, inclusion criteria, and outcome measures. The various staging systems used by the different pediatric and adult cooperative groups are shown in Tables 4.6, 4.7, 4.8, 4.9, 4.10, and 4.11. Therefore, this review and the summary of results in Tables 4.1, 4.2, 4.3, 4.4, and 4.5 should not be seen as a comparison of relative effectiveness. Rather, it is meant to reflect the diversity and common aspects of approaches available globally for the treatment of childhood MGCTs.

Despite the differences in the details, MGCT by its nature lends itself to a trichotomous "functional" classification, which also neatly outlines the clinical and research priorities for each of three risk groups.

In this scheme, a low-risk group can be defined as one where patients can be initially managed with surgical resection alone followed by active surveillance. Adjuvant chemotherapy is reserved only for those patients who recur on surveillance. While first events may increase without the use of upfront chemotherapy, the second event-free survival and overall survival of patients treated on this strategy are unaffected and close to 100 %. The majority of children can thus be spared chemotherapy. It is generally

Table 4.1 Summary of results in pediatric GCT trials, North America

Study	Inclusion	Site/stage	Treatment	Number	EFS	OS	Regimen	Author
POG 9048/CCG 8891 (INT-0106) 1990–1995	NGGCT, low- and intermediate-risk groups	Testis I[a]	Surgery	63	82 % at 6 y	100 %	PEB 4 cycles every 21 days: Cisplatin 20 mg/m²/day×5 days Etoposide 100 mg/m²/day×5 days Bleomycin 15 mg/m² on day 1 If not in CR after 4 cycles, 2 additional cycles given	Schlatter et al. (2003) Rogers et al. (2004)
		Testis II[a]	PEB	17	100 %	100 %		
		Ovary I	PEB	41	95 %	95 %		
		Ovary II	PEB	16	88 %	94 %		
		All int risk	PEB	N=74	95 %	96 %		
POG 9049/ CCG8891 (INT – 0097) 1990–1996	NGGCT and germinomas, high-risk group	Testis III	PEB	8	100 % at 6 y	100 %	PEB 4 cycles every 21 days: Cisplatin 20 mg/m²/day×5 days Etoposide 100 mg/m²/day×5 days Bleomycin 15 mg/m² on day 1 High-dose PEB: Cisplatin 40 mg/m²/day×5 days Etoposide 100 mg/m²/day×5 days Bleomycin 15 mg/m² on day 1 If not in CR after 4 cycles, 2 additional cycles given	Cushing et al. (2004)
			HD-PEB	9	89 %	100 %		
		Testis IV	PEB	22	86 %	91 %		
			HD-PEB	21	91 %	91 %		
		Ovary III	PEB	31	94 %	97 %		
			HD-PEB	27	100 %	100 %		
		Ovary IV	PEB	6	67 %	83 %		
			HD-PEB	10	100 %	100 %		
		EG I-II	PEB	13	85 %	92 %		
			HD-PEB	17	92 %	92 %		
		EG III-IV	PEB					
			HD-PEB					
		Overall	PEB	150	81 %	86 %		
			HD-PEB	149	90 %	92 %		

(continued)

Table 4.1 (continued)

Study	Inclusion	Site/stage	Treatment	Number	EFS	OS	Regimen	Author
AGCT0132 2003–2011	NGGCT, low- and intermediate-risk groups	Testis I	Surgery	80	74 % at 3 y	100 %	Compressed PEB every 21 days for 3 cycles: Cisplatin 33.3 mg/m²/day × 3 days Etoposide 167 mg/m²/day × 3 days Bleomycin 15 mg/m² on day 1 If not in CR after 3 cycles, 3 additional cycles given	Preliminary results
		Ovary I	Surgery	31	52 %	95 %		
		All low risk	PEB compress	N = 105	68 %	99 %		
		Testis II–III	PEB compress	33	93 %	95 %		
		Testis IV	PEB compress	13	52 %			
		Ovary II–III	PEB compress	119	85 %			
		EG I–II	PEB compress	36	100 %			
		All int risk		N = 201	87 %			
AGCT01P1 2004–2008	NGGCT, high-risk group	EG III–IV	C-BEP	18	71 % at 2 y	88 %	C-PEB: Cyclophosphamide 1.2–2.4 g/m² Cisplatin 20 mg/m² × 5 days Etoposide 100 mg/m² × 5 days Bleomycin 15 mg/m² on day 1 If not in CR after 4 cycles, 2 additional cycles given	Preliminary results

Blank fields indicate information not available

Abbreviations: C-PEB cyclophosphamide, cisplatin, etoposide, and bleomycin, *CR* complete response, *EG* extragonadal, *EFS* event-free survival, *HD-PEB* high-dose cisplatin with etoposide and bleomycin, *NGGCTs* non-germinomatous germ cell tumors, *OS* overall survival, *PEB* cisplatin, etoposide, and bleomycin, *y* years

[a]Under age 10 years only

Table 4.2 Pediatric GCT trials, United Kingdom

Study	Inclusion	Site/stage	Treatment	Number	EFS	OS	Regimen	Author
GC I 1979–1987	NGGCT and germinomas, all risk groups, age <16 years	Testis I All others	Surgery PEB[a]	53 33	41/53	100 % 84 %	PEB: Cisplatin 100 mg/m^2 on day 1 Etoposide 120 mg/m^2/day × 3 days Bleomycin 15 mg/m^2 on day 2[b] Given until CR plus 2 additional cycles (median 5 cycles)	Mann et al. (1989)
GC II 1989–1997	NGGCT and germinomas, all risk groups, age <16 years	Testis I Other stage I Stage II Stage III Stage IV All JEB pts	Surgery JEB JEB JEB JEB	51 22 32 35 42 N = 137[c]	40/51 100 % 94 % 85 % 78 % 88 %	100 % 91 %	JEB every 3–4 weeks: Carboplatin 600 mg/m^2 on day 2 Etoposide 120 mg/m^2/day for 3 days Bleomycin 15 mg/m^2 on day 3 Given until CR plus 2 additional cycles (median 5 cycles)	Mann et al. (2000)

Abbreviations: JEB carboplatin, etoposide, and bleomycin
[a]The study included 126 children of whom 78 received chemotherapy with various regimens. Only those who received platinum-based chemotherapy are shown here
[b]Six children were given bleomycin weekly
[c]Six children had unknown stage

agreed that patients with stage I testicular tumors that are completely resected via an inguinal orchiectomy and have normalized tumor markers can be managed with this strategy (Schlatter et al. 2003; Wood et al. 2010). An area of ongoing research and controversy is whether the same strategy can be applied to other patients, including completely resected ovarian or extragonadal tumors.

An intermediate-risk group defines those patients who do require chemotherapy but who have excellent outcomes with current regimens. The research priority for these children focuses on maintaining the high cure rates while minimizing late effects or the burden of treatment.

Lastly, a high-risk group represents patients who have unsatisfactory outcomes on current regimens and for whom further improvements in cure rates are still needed.

4.2.1 The North American Experience

In the 1990s, the North American Pediatric Oncology Group (POG) and the Children's Cancer Group (CCG) together conducted two intergroup studies to determine the optimal management for children with MGCTs (Tables 4.1 and 4.6). The North American studies utilized the PEB regimen as developed in adult studies, but due to concerns regarding pulmonary toxicity in children, they decreased the frequency of administration of bleomycin from once every week to once every 3 weeks. Of note, the reduced frequency of bleomycin was never studied in a comparative manner against the weekly dose administration.

The POG 9048/CCG 8882 study for low- and intermediate-risk groups asked firstly whether stage I testicular NGGCTs in prepubertal boys (under 10 years old) could be successfully treated

Table 4.3 Pediatric GCT trials, German Society of Pediatric Oncology

Study	Inclusion	Site/stage	Treatment	Number	EFS	OS	Regimen	Author
MAHO 82/88/92/98 1982–2001	Testis site: 199 NGGCT, 2 germinoma, 59 teratoma	IA IB–IIB IIC–IV All pts	Surgery PVB PEB	185 75 chemo N=260	97 % DFS at 5 years		PVB for 3 cycles (MAHO98): Cisplatin 20 mg/m^2 for 5 days Vinblastine 3 mg/m^2 for 2 days Bleomycin 15 mg/m^2 for 3 days[a] PEB for 3 cycles (MAHO98): Cisplatin 20 mg/m^2 for 5 days Etoposide 80 mg/m^2 for3 days Bleomycin 15 mg/m^2 for3 days[a] Changed to PEI for incomplete responders	Schmidt et al. (2002), Gobel et al. (2000)
MAKEI 83/86/89/96	Non-testis sites: 293 NGGCT, 114 germinoma, 393 teratoma	Resected T1 All others By histology: Germinoma NGGCT	Surgery alone Chemo	114 293	86 % at 10 years 81 %[b]		MAKEI 83/86: 4PVB+4 PEI MAKEI 89: 3–4 PEB+3–4 VIP MAKEI 96: 4–5 PEI PEI (MAKEI 96): Cisplatin 20 mg/m^2 for 5 days Etoposide 100 mg/m^2 for 3 days Ifosfamide 1,500 mg/m^2 for 5 days	Gobel et al. (2000)

Abbreviations: *DFS* disease-free survival, *PVB* cisplatin, vinblastine and bleomycin, *PEI* cisplatin, etoposide, and ifosfamide, *PEB* cisplatin, etoposide, and bleomycin, *VIP* vinblastine, ifosfamide, and cisplatin
aOmitted in children under 1 year, half dose in children 1–2 years
bReported results include children with intracranial germ cell tumors

Table 4.4 Pediatric GCT trials, French Society of Pediatric Oncology

Study	Inclusion	Site/stage	Treatment	Number	EFS	OS	Regimen	Author
TGM 85	Localized NGGCT[a]	Completely resected	Surgery alone	20			CA/PVB for 3 cycles: Actinomycin 10 μg/kg days 1–5 Cyclophosphamide 300 mg/m² days 1–5 Vinblastine 3 mg/m² days 22–23 Bleomycin 15 mg/m² days 22–23 Cisplatin 100 mg/m² days 24	Baranzelli et al. (1999)
1985–1989		Incompletely resected	CA/PVB	41	CR 90 %	88 % at 3 years		
TGM 90	Localized NGGCT[a]	Completely resected	Surgery alone	34			CA/JVB: Actinomycin 15 μg/kg days 22–24 Cyclophosphamide 500 mg/m² days 22–24 Vinblastine 3 mg/m² days 1–2 Bleomycin 15 mg/m² days 1–2 Carboplatin 400 mg/m² days 3 Given until CR plus 2 additional cycles	Baranzelli et al. (1999)
1990–1994		Incompletely resected	CA/JVB	40	CR 58 %	78 % at 3 years		

Abbreviations: *CA* cyclophosphamide and actinomycin, *JVB* carboplatin, vinblastine, and bleomycin
[a]Children under 1 year old were excluded from published analysis

Table 4.5 Pediatric GCT trials, Brazilian Pediatric Oncology Society

Study	Inclusion	Site/stage	Treatment	Number	EFS	OS	Regimen	Author
GCT-91	52 NGGCT, 14 germinomas, 40 teratomas	Testis I	Surgery	35			PE for 5 cycles: Cisplatin 20 mg/m² x 5 days Etoposide 100 mg/m² × 5 days HPE for 5 cycles: Cisplatin 30 mg/m² × 5 days Etoposide 120 mg/m² × 5 days Changed to VIB if not in CR by 3 cycles	Lopes et al. (2009)
1991–2000		Testis II, Ovary Ic, EG I–II	PE	18		89 % at 5 years		
		All sites III–IV	HPE	53		76 %		

Abbreviations: *HPE* high-dose cisplatin and etoposide, *VIB* vinblastine, ifosfamide, and bleomycin

Table 4.6 Staging system used by Children's Oncology Group GCT trials

Staging of testicular, ovarian, and extragonadal tumors

Testicular

I	Limited to testis, completely resected by high inguinal orchiectomy; no clinical, radiologic, or histologic evidence of disease beyond the testis; tumor markers normal after appropriate half-life decline; patients with normal or unknown markers at diagnosis must have negative ipsilateral retroperitoneal lymph node sampling to confirm stage I disease if radiographic studies demonstrate lymph nodes >2 cm. Patients who have undergone scrotal orchiectomy with high ligation of cord are stage I
II	Transcrotal biopsy; microscopic disease in scrotum or high in spermatic cord (≤5 cm from proximal end). Failure of tumor markers to normalize or decrease with appropriate half-life
III	Retroperitoneal lymph node involvement but no visceral or extra-abdominal involvement. Lymph nodes >4 cm by CT or >2 and >4 cm with biopsy proof
IV	Distant metastases, including liver

Ovarian

I	Limited to ovary (peritoneal evaluation should be negative). No clinical, radiographic, or histologic evidence of disease beyond the ovaries. (Note: presence of gliomatosis peritonei does not result in changing stage I disease to a higher stage.)
II	Microscopic residual; peritoneal evaluation negative. (Note: presence of gliomatosis peritonei does not result in changing stage II disease to a higher stage.) Failure of tumor markers to normalize or decrease with an appropriate half-life
III	Lymph node involvement (metastatic nodule); gross residual or biopsy only; contiguous visceral involvement (omentum, intestine, bladder); peritoneal evaluation positive for malignancy
IV	Distant metastases, including liver

Extragonadal

I	Complete resection at any site; coccygectomy for sacrococcygeal site; negative tumor margins
II	Microscopic residual; lymph nodes negative
III	Lymph node involvement with metastatic disease. Gross residual or biopsy only, retroperitoneal nodes negative or positive
IV	Distant metastases, including liver

Adapted from Children's Oncology Group protocol AGCT0132

with surgical resection and surveillance alone (Schlatter et al. 2003; Rogers et al. 2004). This strategy produced a 6-year event-free survival (EFS) of 82 %. All boys who recurred could be rescued with chemotherapy, and the overall survival of this approach was 100 %.

Secondly, it asked whether children with stage II testicular and stage I–II ovarian GCTs could be treated with standard-dose PEB, with bleomycin given once every 3 weeks. Standard PEB included cisplatin at a dose of 20 mg/m^2/day for 5 days (100 mg/m^2/cycle) for a total of 4 cycles. Children in each group had excellent overall survivals of over 93 % (Billmire et al. 2004; Rogers et al. 2004).

The second intergroup study, POG 9049/CCG 8891, was a randomized trial that investigated whether a twofold cisplatin dose escalation could improve survival in high-risk patients compared to standard-dose PEB (Cushing et al. 2004). The high-risk group for this trial included children with stage III–IV gonadal or stage I–IV extragonadal tumors. The high-dose PEB (HD-PEB) regimen administered cisplatin at a dose of 40 mg/m^2/day for 5 days (200 mg/m^2/cycle). Children who were not in CR after 4 cycles received an additional 2 cycles of the assigned treatment as consolidation. Thus, the cumulative dose of cisplatin in the high-dose arm was 800–1,200 mg/m^2.

With cisplatin dose intensification, EFS improved significantly from 80 to 90 %. OS improved from 86 to 92 % but did not achieve statistical significance. However, the utility of the high-dose PEB strategy was limited by its significant ototoxicity. In the HD-PEB arm, 67 % of children needed hearing aids, as opposed to about 10 % in the standard-dose PEB arm (Cushing et al. 2004; Li et al. 2004).

The results of the first two intergroup studies allowed refining the risk stratification for the next generation of studies by the Children's Oncology Group (COG), AGCT0132 and AGCT01P1. The newly stratified low-, intermediate- and high-risk groups followed the aforementioned functional classification scheme (Marina 2006).

The low-risk group now included stage I testicular and stage I ovarian GCTs. It was hypothesized that stage I ovarian GCTs might also be successfully managed with surgery and surveillance alone, with PEB chemotherapy reserved for recurrences. Emerging evidence

Table 4.7 GCT staging system used by UKCCSG trial

Stage	Site of origin of tumor			
	Testis	Ovary	Uterus, vagina, prostate of sacrococcygeal region	Abdomen, retroperitoneum, thorax, or others
I	Tumor confined to testis	Tumor confined to ovary	Tumor confined to organ/site of origin	Tumor confined to the site of origin and resectable
II	Tumor spread limited to retroperitoneal lymph nodes	Tumor spread limited to pelvis	Tumor spread limited to the pelvis	Local spread
III	Tumor spread limited to retroperitoneal and/or mediastinal/supraclavicular nodes	Tumor spread limited to pelvis and abdomen, excluding liver	Tumor spread limited to pelvis and abdomen, excluding liver	Extensive spread confined to one side of the diaphragm, excluding liver
IV	Tumor spread to liver, lung, bone, brain, etc.	Tumor spread to liver or beyond abdominal cavity	Tumor spread to the liver or beyond the abdominal cavity	Tumor spread to the liver, to both sides of the diaphragm, and/or to bones, bone marrow, or brain

Adapted from Mann et al. (2000)

Table 4.8 Staging system used by German MAHO studies for testicular GCT

Lugano staging system	
I	No evidence of metastatic spread
IA	Tumor limited to the testis and its appendages
IB	Tumor with infiltration of the epididymis or in cryptorchidism
IC	Tumor infiltrates scrotum or was resected transcrotally or rose after inguinal or scrotal surgery
IX	Extent of the primary tumor cannot be determined
II	Infradiaphragmatic lymph node metastases
IIA	All lymph nodes ≤2 cm
IIB	At least 1 lymph node 2–5 cm
IIC	Retroperitoneal lymph nodes >5 cm
IID	Palpable abdominal tumor or fixed inguinal tumor
III	Mediastinal or supraclavicular lymph node metastases, distant metastases
IIIA	Mediastinal and/or supraclavicular lymph node metastases (no distant metastases)
IIIB	Lung metastases
	Minimal pulmonary disease: less than 5 nodules per lung <2 cm
	Advanced pulmonary disease: more than 5 nodules per lung or 1 pulmonary nodule >2 cm or pleural effusion
IIIC	Extrapulmonary distant metastases
IIID	Persistent tumor markers after orchiectomy without detectable metastases

Adapted from Cavalli et al. (1980)

from observational studies had suggested that surveillance might also be a safe and feasible strategy for completely resected ovarian GCTs (Dark et al. 1997; Mitchell et al. 1999; Baranzelli et al. 2000; Gobel et al. 2000, 2003; Mann et al. 2000; Palenzuela et al. 2008; Patterson et al. 2008). There were 31 girls with stage I ovarian non-germinomatous germ cell tumors (NGGCTs) enrolled on the AGCT0132 study, of whom 12 had recurrent disease and were then treated with PEB. One patient died, but she had shown only a transient response to PEB chemotherapy. The 3-year EFS was 52 % and OS was 95 % (Frazier et al. 2010). Thus, ~50 % of patients were spared all exposure to chemotherapy, and the overall survival outcome may have been no different when compared to a strategy of standard adjuvant chemotherapy. These results suggest that surveillance may be an acceptable option for stage I ovarian MGCTs. The higher proportion of recurrences compared to stage I testicular MGCTs may not be due to an intrinsic biologic difference but rather may reflect the increased difficulty of accurately staging the often larger, intra-abdominal ovarian tumors.

The intermediate-risk group now included stage II–IV testicular, stage II–III ovarian, and stage I–II extragonadal tumors. Note that the intermediate-risk group was expanded to include

Table 4.9 Staging system used by FIGO for ovarian tumors

Stage	
I	Growth limited to the ovaries
Ia	Growth limited to one ovary; no ascites present containing malignant cells. No tumor on the external surface; capsule intact
Ib	Growth limited to both ovaries; no ascites present containing malignant cells. No tumor on the external surface; capsule intact
Ic	Tumor either stage Ia or Ib but with tumor on surface of one or both ovaries, or with capsule ruptured, or with ascites present containing malignant cells, or with positive peritoneal washings
II	Growth involving one or both ovaries with pelvic extension
IIa	Extension and/or metastases to the uterus and/or tubes
IIb	Extension to other pelvic tissues
IIc	Tumor either stage IIa or IIb but with tumor on surface or one or both ovaries, or with capsule(s) ruptured, or with ascites present containing malignant cells, or with positive peritoneal washings
III	Tumor involving one or both ovaries with histologically confirmed peritoneal implants outside the pelvis and/or positive regional lymph nodes. Superficial liver metastases equals stage III. Tumor is limited to the true pelvis but with histologically proven malignant extension to small bowel or omentum
IIIa	Tumor grossly limited to the true pelvis, with negative nodes, but with histologically confirmed microscopic seeding of abdominal peritoneal surfaces or histologic proven extension to small bowel or mesentery
IIIb	Tumor of one or both ovaries with histologically confirmed implants, peritoneal metastasis of abdominal peritoneal surfaces, none exceeding 2 cm in diameter; nodes are negative
IIIc	Peritoneal metastasis beyond the pelvis more than 2 cm in diameter and/or positive regional lymph nodes
IV	Growth involving one or both ovaries with distant metastases. If pleural effusion is present, there must be positive cytology to allot a case to stage IV. Parenchymal liver metastasis equals stage IV

Adapted from FIGO Committee on Gynecologic Oncology (2009)

additional stages based on the excellent outcomes with even standard-dose PEB on the intergroup studies. As a group, the intermediate-risk patients

experienced an EFS over 92 %. The research priority for these children therefore focused on maintaining the high cure rates while minimizing toxicity. AGCT0132 investigated whether 3 cycles of PEB (compressed from 5 days into a 3 day regimen) could achieve equivalent outcomes to the 4 cycles used previously. This hypothesis was based on the experience of two adult randomized trials in good-prognosis metastatic testicular GCTs, where 3 cycles of PEB with weekly bleomycin were non-inferior to 4 cycles (Einhorn et al. 1989; Culine et al. 1997; Saxman et al. 1998; De Wit et al. 2001). Although published results are not yet available, preliminary analysis revealed that 3 cycles was sufficient to maintain an EFS of 92 %, except for boys with stage IV testicular tumors and girls with stage III ovarian tumors. Thus, for most intermediate-risk patients, 3 cycles of PEB (when bleomycin is used every 3 weeks) are likely sufficient.

The high-risk group, which was narrowed to include only stage III–IV extragonadal tumors, represented the group for which further improvements in cure rates were still needed. Pursuing this increased cure has followed two general strategies in cooperative group studies: discovering ways (such as effective otoprotectants) to allow dose-intensive regimens to be used with less toxicity or adding new agents to the PEB combination.

The first strategy, utilizing otoprotectants to allow the delivery of high-dose PEB, has been an area of active research. The COG study P9749 investigated the addition of amifostine to high-dose PEB but found it could not reduce the high rates of ototoxicity (Marina et al. 2005). Amifostine was administered at a dose of 825 mg/m^2/day for 5 days per cycle to 25 children, and the rate of significant (grades 2–4) ototoxicity was 75 %, identical to the previous study of high-dose PEB (Cushing et al. 2004). In contrast, a study of amifostine in 97 children with average-risk medulloblastoma did note a reduction in the rate of ototoxicity (from 37 to 15 %), where amifostine was administered as a total dose (over 2 boluses) of 1,200 mg/m^2 in 1 day (Fouladi et al. 2008). The COG is currently also investigating the use

Table 4.10 Staging of testicular tumors used by American Joint Committee on Cancer

A. TNM staging	Unit	Value
Primary tumor (T)[a]	pTX	Primary tumor cannot be assessed
	pT0	No evidence of primary tumor (e.g., histologic scar in testis)
	pTis	Intratubular germ cell neoplasia (carcinoma in situ)
	pT1	Tumor limited to the testis and epididymis without vascular/lymphatic invasion; tumor may invade into the tunica albuginea but not the tunica vaginalis
	pT2	Tumor limited to the testis and epididymis with vascular/lymphatic invasion or tumor extending through the tunica albuginea with involvement of the tunica vaginalis
	pT3	Tumor invades the spermatic cord with or without vascular/lymphatic invasion
	pT4	Tumor invades the scrotum with or without vascular/lymphatic invasion
Regional lymph nodes (N)	NX	Regional lymph nodes cannot be assessed
	N0	No regional lymph node metastasis
	N1	Metastasis with a lymph node mass 2 cm or less in greatest dimension or multiple lymph nodes, none more than 2 cm in greatest dimension
	N2	Metastasis with a lymph node mass more than 2 cm but not more than 5 cm in greatest dimension or multiple lymph nodes or any mass greater than 2 cm but not more than 5 cm in greatest dimension
	N3	Metastasis with a lymph node mass more than 5 cm in greatest dimension
Distant metastasis (M)	M0	No distant metastasis
	M1	Distant metastasis
	M1a	Nonregional nodal or pulmonary metastasis
	M1b	Distant metastasis other than to nonregional lymph nodes and lung
Serum tumor markers (S)	SX	Marker studies not available or not performed
	S0	Marker study levels within normal limits
	S1	LDH <1.5 times normal (N) *and* hCG (mIu/mL) <5,000 *and* AFP (ng/mL) <1,000
	S2	LDH 1.5–10×N *or* hCG (mIu/mL) 5,000–50,000 *or* AFP (ng/mL) 1,000–10,000
	S3	LDH >10×N *or* hCG (mIu/mL) >50,000 *or* AFP (ng/mL) >10,000

B. Anatomic stage/prognostic groups				
Group	T	N	M	S
Stage 0	pTis	N0	M0	S0
Stage I	pT1–4	N0	M0	SX
Stage IA	pT1	N0	M0	S0
Stage IB	pT2	N0	M0	S0
	pT3	N0	M0	S0
	pT4	N0	M0	S0
Stage IS	Any pT/Tx	N0	M0	S1–3 (measured post orchiectomy)
Stage II	Any pT/Tx	N1-3	M0	SX

(continued)

Table 4.10 (continued)

Stage IIA	Any pT/Tx	N1	M0	S0
	Any pT/Tx	N1	M0	S1
Stage IIB	Any pT/Tx	N2	M0	S0
	Any pT/Tx	N2	M0	S1
Stage IIC	Any pT/Tx	N3	M0	S0
	Any pT/Tx	N3	M0	S1
Stage III	Any pT/Tx	Any N	M1	SX
Stage IIIA	Any pT/Tx	Any N	M1a	S0
	Any pT/Tx	Any N	M1a	S1
Stage IIIB	Any pT/Tx	N1-3	M0	S2
	Any pT/Tx	Any N	M1a	S2
Stage IIIC	Any pT/Tx	N1-3	M0	S3
	Any pT/Tx	Any N	M1a	S3
	Any pT/Tx	Any N	M1b	Any S

Adapted from American Joint Committee on Cancer (2010)

[a]The extent of primary tumor is usually classified after radical orchiectomy, and, for this reason, a pathologic (p) stage is assigned. Except for pTis and pT4, extent of primary tumor is classified by radical orchiectomy. TX may be used for other categories in the absence of radical orchiectomy

Table 4.11 Risk stratification system for metastatic testicular GCTs used by the International Germ Cell Cancer Collaborative Group

Histology	Prognostic category	Clinical factors
Nonseminoma germ cell tumor	Good	Testes/retroperitoneal primary and no nonpulmonary visceral metastases and good markers: AFP <1,000 ng/mL and HCG <5,000 IU/L and LDH <1.5×upper limit of normal (ULN)
	Intermediate	Testes/retroperitoneal primary and no nonpulmonary visceral metastases and intermediate markers: AFP ≥1,000 ng/mL and ≤10,000 ng/mL or HCG ≥5,000 IU/L and ≤50,000 ng/mL or LDH ≥1.5×ULN and ≤10×ULN
	Poor	Mediastinal primary or nonpulmonary visceral metastases or poor markers with any of the following: AFP >10,000 ng/mL or HCG >50,000 IU/L or LDH >10×ULN
Seminoma	Good	Any primary site and no nonpulmonary visceral metastases and normal AFP, any HCG, any LDH
	Intermediate	Any primary site and nonpulmonary visceral metastases and normal AFP, any HCG, any LDH

Adapted from International Prognostic Factors Study Group (1997)

of sodium thiosulfate for the same objective of reducing cisplatin ototoxicity in the study ACCL0431 (Freyer et al. 2009).

The second strategy was tested by AGCT01P1, a pilot COG study to establish the maximum tolerated dose and toxicity profile of cyclophosphamide combined with PEB (C-PEB). The study noted 5 relapses among 18 evaluable children with stage III–IV extragonadal tumors but was not powered to compare efficacy. Randomized trials in adults have not yet succeeded in finding any regimen that confers an advantage over standard PEB (Samson et al. 1984; Ozols et al. 1988; Lewis et al. 1991; Nichols et al. 1991; Kaye et al. 1998; Christian et al. 2003; Fossa et al. 2005; Rexer 2005). There is also no evidence to date that the use of high-dose chemotherapy with autologous stem cell transplant in adults is superior to PEB for patients with poor-prognosis metastatic testicular MGCTs (Droz et al. 2007; Motzer et al. 2007), although research is ongoing. It is likely that further progress for the high-risk group of patients will be achieved only through a better understanding of biology, identification of molecular targets, and the development of new therapeutic agents.

4.2.2 The United Kingdom Experience

The use of cisplatin carries with it a significant burden of toxicity, including emetogenicity, ototoxicity, and nephrotoxicity. Therefore, one potential method of attempting to reduce the toxicity of treatment was to find a platinum agent with a more favorable spectrum of adverse effects. To this end, several randomized controlled trials (RCTs) in adults with metastatic testicular MGCTs sought to compare cisplatin-containing regimens to carboplatin-containing regimens (Bajorin et al. 1993; Tjulandin et al. 1993; Bosl and Bajorin 1994; Bokemeyer et al. 1996; Horwich et al. 1997, 2000; Bokemeyer et al. 2004). These trials in general observed an inferior effect of carboplatin regimens for this population. Adult oncologists thus accepted cisplatin regimens as their standard of care.

However, many pediatric oncologists remained concerned about the risk-benefit balance of cisplatin chemotherapy. The significance and consequences of ototoxicity are magnified in children compared to adults. Permanent, high-frequency, bilateral hearing loss is reported in 10–20 % of children who receive standard-dose cisplatin and up to 75 % of children who receive high-dose cisplatin (Bertolini et al. 2004; Cushing et al. 2004; Li et al. 2004; Marina et al. 2005; Chang and Chinosornvatana 2010; Neuwelt and

Brock 2010). Loss of auditory acuity in young children leads to differences in speech-language development and literacy. In older children and adolescents, hearing loss impacts educational achievement, social-emotional development, and quality of life. Platinum is still detectable in plasma and urine up to 20 years after administration of cisplatin (Gerl and Schierl 2000; Gietema et al. 2000). Adult survivors treated with chemotherapy for testicular cancer are at increased risk of second malignancies and cardiovascular disease (Belt-Dusebout et al. 2007; Travis et al. 2010). Finding ways to mitigate the long-term toxicity of cisplatin-based chemotherapy has therefore long been an important research objective for pediatric oncology collaborative groups, as survivors of childhood MGCT can potentially have 70 or more years of remaining life.

In 1989, at the same time as similar studies were being conducted in adults, the United Kingdom Children's Cancer Study Group (UKCCSG), now the Children's Cancer and Leukaemia Group (CCLG), substituted carboplatin for cisplatin in its second germ cell tumor study (GCII) in an effort to reduce the rates of ototoxicity and nephrotoxicity in childhood survivors while maintaining high cure rates (Mann et al. 1989, 1998, 2000; Huddart et al. 1990; Pinkerton et al. 1990). The UK JEB regimen also administered bleomycin once every 3 weeks, due to the observed pulmonary toxicity of weekly bleomycin in young children on a prior study (Mann et al. 1989).

The GCII trial (Tables 4.2 and 4.7) used carboplatin at a dose of 600 mg/m² or by formula to achieve a target area under the plasma concentration-versus-time curve (AUC) of 6 mg/mL/min. In practice, about 75 % of patients received the meter-squared dosing, and retrospectively it has been appreciated that this dose corresponded to a mean AUC of 7.9 mg/mL/min. Courses were given every 3 weeks or upon hematologic recovery, until remission plus two additional courses, with a median of five courses administered. For 137 children treated with JEB, the 5-year EFS was 87.8 % and OS was 90.9 % (Mann et al. 2000). These results are comparable with the reported outcomes using PEB in pediatric MGCTs. There was no sensorineural deafness and no reports of significant nephrotoxicity. The principal complication was myelotoxicity, but there were no cytopenia-related deaths. It is worth noting that carboplatin here was administered at a higher dose than in each of the adult trials that have inferred its inferiority to cisplatin, which used carboplatin doses ranging from 350 to 500 mg/m² (Bajorin et al. 1993; Tjulandin et al. 1993; Bokemeyer et al. 1996) or an AUC of 5 mg/mL/min (Horwich et al. 1997; Bokemeyer et al. 2004).

It is possible that the good results observed in this study related to the higher dose of carboplatin, the larger number of courses, or an inherent biological difference of childhood MGCT. The results of this study suggested that carboplatin could be substituted for cisplatin in childhood MGCT as long as the dose is sufficient. The UKCCSG and several other institutions internationally have adopted JEB as the standard regimen for the treatment of newly diagnosed pediatric MGCTs.

The current GC III study continues the use of the JEB regimen. It stratifies patients to one of three risk groups. Those in the low-risk group with normal tumor markers undergo observation alone. Intermediate-risk patients receive JEB chemotherapy for 4 cycles, and high-risk patients receive JEB for 6 cycles. Salvage therapy after JEB consists of vincristine, ifosfamide, and cisplatin (VeIP) for 6 cycles.

4.2.3 The German Experience

The German Society of Pediatric Oncology and Hematology (GPOH) has conducted several cooperative studies of GCT treatment under the titles of MAHO for testicular and MAKEI for non-testicular site tumors (Table 4.3). The GPOH studies utilized the Lugano staging system (Table 4.8) for testicular tumors (Cavalli et al. 1980), the FIGO system (Table 4.9) for ovarian tumors (Cannistra 1993; FIGO Committee on Gynecologic Oncology 2009), and the TNM soft tissue sarcoma staging system for non-testicular tumors (Russell et al. 1977). Between 1982 and 2009, a total of 2,077

patients with both intracranial and extracranial germ cell tumors have been prospectively enrolled onto these studies, with an overall survival of 88 % (Gobel et al. 2010).

The MAHO studies in general managed only stage IA (limited to the testis) teratomas or pure yolk sac tumors with a resection and surveillance strategy. All other patients received adjuvant chemotherapy. The MAHO 82, 88 and 92 studies utilized four courses of PVB, with a switch to PEI for those with incomplete response after 2 cycles (Haas et al. 1995; Gobel et al. 2000; Schmidt et al. 2002). The more recent MAHO 98 protocol administered 2–3 cycles of PVB for patients who are stage IB (tumor infiltrating epididymis) to IIB (at least one lymph node 2–5 cm) and 3 cycles of PEB to patients who are stage IIC or above. PEI is given to those with incomplete response. Between 1982 and 2001, 260 patients were treated on MAHO protocols, of whom only 75 required chemotherapy. The disease-free survival at a median follow-up of 5 years was 97 % for all patients.

The MAKEI series of studies utilized a risk-stratified, multimodal approach to the management of non-testicular GCTs, and chemotherapy protocols evolved in each successive protocol, as summarized in Table 4.3. In general, subsequent studies tended to a reduction of cumulative doses of chemotherapy while maintaining favorable outcomes (Gobel et al. 2000). A stepwise reduction from 8 cycles on the 83/86 study to 6–8 on the 89 study and 4–5 cycles on the 96 study was achieved. The 10-year EFS pooled over all four studies was 81 % for NGGCTs and 86 % for germinomas (of note, this reported outcome includes intracranial tumors) (Gobel et al. 2000).

Another trend was the sequential reduction in exposure to bleomycin. The 83/86 studies applied bleomycin at a dose of 15 mg/m²/day for 3 days per cycle over 4 cycles, for a cumulative bleomycin dose of 180 mg/m². The 89 protocol eliminated bleomycin for children under 1 year old and used a half dose for children between 1 and 2 years of age. Subsequently, the 96 study eliminated the use of bleomycin altogether in favor of a PEI regimen, prompted by the observation of severe or lethal bleomycin-associated pulmonary toxicity, especially in young children.

As a result, the recent MAKEI 96 study utilized a surveillance strategy for completely resected stage T1 tumors (less than 5 cm in size), 2–3 cycles of PE for completely resected stage T2 tumors (5 cm or greater), and 4–5 cycles of PEI for incompletely resected tumors.

4.2.4 The French Experience

The French Society of Pediatric Oncology (SFOP) conducted two studies of localized childhood MGCT between 1985 and 1994. Together, the TGM 85 and TGM 90 studies comprised of 152 children with newly diagnosed, localized, NGGCTs (Baranzelli et al. 1993, 1999, 2000). Children with metastatic disease were treated on separate, more intensive regimens.

The French protocols were unique in that they offered a surveillance strategy without initial systemic chemotherapy to all completely resected localized tumors with normal tumor markers, including those of non-testicular sites. Thus, only 81 patients across both studies received initial chemotherapy. Children under 1 year old were excluded from the published analysis in order to allow determination of the prognostic importance of AFP levels. The combined 3-year failure-free survival rate was 68 %, and the overall survival rate was 87 % for these 81 patients.

Both the TGM 85 and TGM 90 used actinomycin, cyclophosphamide, bleomycin, and vinblastine in similar doses, although the schedule of administration varied slightly. The primary difference between the two protocols was the choice of platinum agent. The TGM 85 used cisplatin at a dose of 100 mg/m²/dose, while the TGM 90 study substituted carboplatin at a dose of 400 mg/m²/dose. Salvage therapy included etoposide and ifosfamide, with or without cisplatin.

The CR rate was 90 % in the TGM 85 protocol. In the TGM 90 protocol, the CR rate after first-line treatment was only 58 %. The overall CCR rate was again 90 % as most patients could achieve CR after switching to second-line therapy. Nevertheless, the OS in TGM 85 was 88 % compared with 78 % in TGM 90. Thus, the TGM studies demonstrated a lower response rate and survival rate of NGGCTs to carboplatin relative to cisplatin in regimens that were otherwise similar. However, in comparison to the GC II study, the TGM 90 study a used a lower dose of carboplatin per cycle, and as sequences of JVB alternated with sequences of AC making each cycle 6 weeks long, the platinum dose intensity was lower as well.

The recent French protocol used cisplatin and maintained platinum dose intensity by using alternating cycles of PE and PI.

4.2.5 The Brazilian Experience

In 1988, a consortium of Brazilian institutions was formed to develop uniform management standards for the treatment of childhood GCT in the country. The first Brazilian germ cell tumor protocol, GCT-91, enrolled 106 assessable patients with GCTs of all histologies between 1991 and 2000, of whom 71 patients received chemotherapy (Lopes et al. 1995, 2009). The protocol utilized a two-agent regimen of cisplatin and etoposide (PE), based on the reported effectiveness in adults and the concern of pulmonary toxicity of bleomycin in young children. PE was given for 5 cycles. The dose of chemotherapy was tailored to the risk group. Twenty-two patients not in CR after 3 cycles of PE were changed to an alternate regimen of vinblastine, ifosfamide, and bleomycin (VIB). Overall, the cumulative doses of chemotherapy were somewhat higher than corresponding North American trials. The 5-year OS for intermediate- and high-risk groups were 88.9 and 75.5 %, respectively. An improvement in survival compared to historical outcomes was seen and could be attributed in part to the establishment of a collaborative national network of centers and a standardized protocol. The authors concluded that the 2-drug regimen was feasible for IR patients and some HR patients.

The second Brazilian germ cell tumor protocol, GCT-99, adapted the COG staging (Table 4.6) and risk-stratification system (Lopes et al. 2010). A total of 533 patients were registered. Children in the intermediate-risk group were treated with 4 cycles of PE, instead of the previous 5 cycles, administered over a compressed 3-day schedule. If patients were not in CR by the third cycle, then 3 cycles of VIB were given. Of 123 children in the intermediate-risk group, the reported 3-year EFS is over 86 %, which compares favorably to the historical control. Final published results are awaited.

4.2.6 Other Experiences

As a disease that is managed primarily by surgery and a short course of chemotherapy, with minimal to no role for radiation therapy or specialized imaging, MGCTs can be effectively treated in low-income countries. An early report from India reported a 3-year EFS of 80 % for 27 children treated with PEB (Nair et al. 1994). A recent report from South Africa described the management of children using surgery and JEB chemotherapy. Seventeen of 19 children were successfully treated, 2 children with advanced disease died, and there was no non-relapse mortality (Hendricks et al. 2011). The JEB regimen was preferred by the treating physicians due to lower costs and higher compliance with an outpatient-based regimen. Such reports show that outcomes in low-income countries are comparable to those of high-income countries and that MGCT can be considered a paradigm of malignancies that may be successfully treated even in settings with limited resources.

4.3   Management of Germinomas and Teratomas

The pediatric trials conducted to date have not stratified treatment based on the histologic categories of germinomatous versus non-germinomatous germ cell tumors. Where germinomas (also referred to as dysgerminomas in the ovary or seminomas in the testis) have been included, they have represented a small minority of the tumor types and have been treated in the same fashion as NGGCTs. Thus,

while germinomas have shown a relatively better outcome compared to NGGCTs where reported (Gobel et al. 2000; Mann et al. 2000), separate treatment recommendations in children cannot presently be inferred.

The discussions thus far have been of the treatment of germ cell tumors of malignant histology. In contrast, mature teratomas (MT) are benign tumors composed of well-differentiated tissues. They are not chemosensitive, and complete surgical resection is an adequate treatment. Immature teratomas (IT) contain varying amounts of immature tissue and can be graded in order of increasing quantity of immature neuroglial tissue as grades 0–3. They are sometimes considered to be tumors of intermediate malignant potential. For some time, it was uncertain whether children with higher grades of IT should receive adjuvant chemotherapy. A study of 58 women with ITs of the ovary treated with surgical resection had found a relapse rate of 70 % among those with grade 3 tumors, and the investigators recommended adjuvant chemotherapy for these patients (Norris et al. 1976).

Subsequent pediatric studies, however, have helped shed some light on the optimal approach to immature teratomas. An intergroup study of the POG and CCG treated 73 children with complete resection (stage I disease) followed by surveillance alone (Marina et al. 1999). Twenty-one (29 %) children had grade 3 tumors. Only five children had recurrences that were then treated with chemotherapy. The overall 3-year EFS was 93 %. Twenty-three patients were found to have foci of malignant tissue on central review, but even among these children only four experienced recurrence after complete resection. All patients were alive at a median follow-up of 35 months, and one patient was alive with disease. The study concluded that observation after resection of ITs was a feasible strategy.

Secondly, the UKCCSG reported their experience of 124 children with immature teratomas (Mann et al. 2008). Sixty-nine (56 %) had grade 3 tumors. The 5-year EFS was 86 % and OS was 95 %. However, 40 % of patients with IT had incomplete surgical resection. Among children who had complete resection, the EFS was 97 %

and OS was 99 %. Six children died, but all deaths were due to surgical or neonatal complications. Incomplete resection and grade 3 IT were associated with worse EFS in multivariable analysis. Recurrences were successfully treated with further surgery or JEB chemotherapy. The study concluded that surgery and observation should remain the mainstay of treatment for children with both mature and immature teratomas.

Conclusions

The experience of the past three decades has led to significant achievements in the management of germ cell tumors in children, but many challenges remain. Future research will focus on refining the treatment strategies for each of the three functional risk groups. Studies will identify ways to predict which patients, in addition to those with stage I testicular disease, can have a high probability of cure after surgical resection alone, thus sparing unnecessary exposure to chemotherapy. For intermediate-risk patients, efforts to identify the most efficacious and minimally toxic regimens will continue. Potential research questions can include the relative risk-benefit ratio of cisplatin versus carboplatin and of regimens that do or do not contain bleomycin or ifosfamide. For high-risk patients, and for those with recurrent or refractory disease, advances will come from new insights in the laboratory and the development of biologically driven innovative therapies (see Chap. 2 for a review of the biology of germ cell tumors).

Each of these advances will be fostered by the development of international collaborative initiatives (Rodriguez-Galindo 2010). As a rare tumor, childhood MGCT presents a small sample size, which limits the ability to conduct effective clinical trials, especially RCTs. The diversity of treatment approaches that have evolved globally can be advantageous in that they may allow clinicians a multitude of options to tailor treatment plans to the specific needs of their individual patients or the specific conditions of their health-care environment. However, the use of different systems for staging, risk grouping, and measurement

of outcome and toxicity also serves to hinder the comparability of results across the pediatric oncology groups.

In adult testicular disease, the creation and analyses of a large, international pooled databases led to the development of unified staging (Table 4.10) and risk classification (Table 4.11) systems (International Prognostic Factors Study Group 1997; Kollmannsberger et al. 2000; Sonneveld et al. 2001; American Joint Committee on Cancer 2010; International Prognostic Factors Study Group 2010). A similar effort has been initiated by some pediatric oncology groups. The COG and the UKCCSG have recently formed a pooled patient database of all their respective trials data, entitled the Malignant Germ Cell Tumors International Collaboration (MaGIC) database, and are using it to determine optimum risk group classifications and treatment approaches. Such efforts can foster the development of evidence-based intergroup consensus, the completion of retrospective comparisons to identify the best elements of each approach, and thereafter the prospective conduct of large-scale international collaborative clinical trials. Together, these advances will allow us to reach ever closer to the goals of curing all children with malignant germ cell tumors and doing so with the least possible late effects.

References

American Joint Committee on Cancer (2010) The AJCC cancer staging manual. Springer, New York

Bajorin DF, Sarosdy MF, Pfister DG et al (1993) Randomized trial of etoposide and cisplatin versus etoposide and carboplatin in patients with good-risk germ cell tumors: a multiinstitutional study. J Clin Oncol 11:598–606

Baranzelli MC, Bouffet E, Quintana E et al (2000) Non-seminomatous ovarian germ cell tumours in children. Eur J Cancer 36:376–383

Baranzelli MC, Flamant F, De Lumley L et al (1993) Treatment of non-metastatic, non-seminomatous malignant germ-cell tumours in childhood: experience of the 'Societe Francaise d'Oncologie Pediatrique' MGCT 1985–1989 study. Med Pediatr Oncol 21(6): 395–401

Baranzelli MC, Kramar A, Bouffet E et al (1999) Prognostic factors in children with localized malignant nonseminomatous germ cell tumors. J Clin Oncol 17(4):1212–1218

Belt-Dusebout AW, De Wit R, Gietema JA et al (2007) Treatment-specific risk of second malignancies and cardiovascular disease in 5-year survivors of testicular cancer. J Clin Oncol 25:4370–4378

Bertolini P, Lassalle M, Mercier G et al (2004) Platinum compound-related ototoxicity in children: long-term follow-up reveals continuous worsening of hearing loss. J Pediatr Hematol Oncol 26:649–655

Billmire D, Vinocur C, Rescorla F et al (2004) Outcome and staging evaluation in malignant germ cell tumors of the ovary in children and adolescents: an intergroup study. J Pediatr Surg 39:424–429

Bokemeyer C, Kohrmann O, Tischler J et al (1996) A randomized trial of cisplatin, etoposide and bleomycin (PEB) versus carboplatin, etoposide and bleomycin (CEB) for patients with 'good-risk' metastatic non-seminomatous germ cell tumors. Ann Oncol 7:1015–1021

Bokemeyer C, Kollmannsberger C, Stenning S et al (2004) Metastatic seminoma treated with either single agent carboplatin or cisplatin-based combination chemotherapy: a pooled analysis of two randomised trials. Br J Cancer 91:683–687

Bosl GJ, Bajorin DF (1994) Etoposide plus carboplatin or cisplatin in good-risk patients with germ cell tumors: a randomized comparison. Semin Oncol 21:61–64

Brodeur GM, Howarth CB, Pratt CB et al (1981) Malignant germ cell tumors in 57 children and adolescents. Cancer 48:1890–1898

Cannistra SA (1993) Cancer of the ovary. N Engl J Med 329:1550

Cavalli F, Monfardini S, Pizzocaro G (1980) Report on the international workshop on staging and treatment of testicular cancer. Eur J Cancer 16:1367–1372

Chang KW, Chinosornvatana N (2010) Practical grading system for evaluating cisplatin ototoxicity in children. J Clin Oncol 28:1788–1795

Christian JA, Huddart RA, Norman A et al (2003) Intensive induction chemotherapy with CBOP/BEP in patients with poor prognosis germ cell tumors. J Clin Oncol 21:871–877

Culine S, Theodore C, Terrier-Lacombe MJ et al (1997) Are 3 cycles of bleomycin, etoposide and cisplatin or 4 cycles of etoposide and cisplatin equivalent optimal regimens for patients with good risk metastatic germ cell tumors of the testis? The need for a randomized trial. J Urol 157(3):855–859

Cushing B, Giller R, Cullen J et al (2004) Randomized comparison of combination chemotherapy with etoposide, bleomycin, and either high-dose or standard-dose cisplatin in children and adolescents with high-risk malignant germ cell tumors: a Pediatric Intergroup Study – Pediatric Oncology Group 9049 and Children's Cancer Group 888. J Clin Oncol 22:2691–2700

Dark GG, Bower M, Newlands ES et al (1997) Surveillance policy for stage I ovarian germ cell tumors. J Clin Oncol 15:620–624

De Wit R, Roberts JT, Wilkinson PM et al (2001) Equivalence of three or four cycles of bleomycin, etoposide, and cisplatin chemotherapy and of a 3- or 5-day schedule in good-prognosis germ cell cancer: a randomized study of the European Organization for Research and Treatment of Cancer Genitourinary Tract Cancer Cooperative Group and the Medical Research Council. J Clin Oncol 19(6):1629–1640

Droz JP, Kramar A, Biron P et al (2007) Failure of high-dose cyclophosphamide and etoposide combined with double-dose cisplatin and bone marrow support in patients with high-volume metastatic nonseminomatous germ-cell tumours: mature results of a randomised trial. Eur Urol 51(3):739–748

Einhorn L (2002) Curing metastatic testicular cancer. Proc Natl Acad Sci U S A 99:4592–4595

Einhorn LH, Donohue JP (1977) Chemotherapy for disseminated testicular cancer. Urol Clin North Am 4:407–426

Einhorn LH, Williams SD, Loehrer PJ et al (1989) Evaluation of optimal duration of chemotherapy in favorable-prognosis disseminated germ cell tumors: a Southeastern Cancer Study Group Protocol. J Clin Oncol 7(3):387–391

FIGO Committee on Gynecologic Oncology (2009) Current FIGO staging for cancer of the vagina, fallopian tube, ovary, and gestational trophoblastic neoplasia. Int J Gynecol Obstet 105:3–4

Fossa SD, Paluchowska B, Horwich A et al (2005) Intensive induction chemotherapy with C-BOP/BEP for intermediate- and poor-risk metastatic germ cell tumours (EORTC trial 30948). Br J Cancer 93:1209–1214

Fouladi M, Chintagumpala M, Ashley D et al (2008) Amifostine protects against cisplatin-induced ototoxicity in children with average-risk medulloblastoma. J Clin Oncol 26:3749–3755

Frazier AL, Billmire D, Schlatter M et al (2010) Stage I germ cell tumors: outcome of a watch and wait strategy [abstract]. Pediatr Blood Cancer 55:805

Freyer DR, Sung L, Reaman GH et al (2009) Prevention of hearing loss in children receiving cisplatin chemotherapy. J Clin Oncol 27:317–318; author reply 318–319

Gerl A, Schierl R (2000) Urinary excretion of platinum in chemotherapy-treated long-term survivors of testicular cancer. Acta Oncol 39:519–522

Gietema JA, Meinardi MT, Messerschmidt J et al (2000) Circulating plasma platinum more than 10 years after cisplatin treatment for testicular cancer. Lancet 355:1075–1076

Gobel U, Calaminus G, Koch S et al (2003) Is a watch and wait treatment possible in all malignant ovarian germ cell tumors (MOGCTs) FIGO 1?: results of the cooperative MAKEI trials [abstract]. Med Pediatr Oncol 41:262

Gobel U, Schneider DT, Calaminus G et al (2000) Germ-cell tumors in childhood and adolescence. GPOH MAKEI and the MAHO study groups. Ann Oncol 11:263–271

Gobel U, Schneider DT, Teske C et al (2010) Brain metastases in children and adolescents with extracranial germ cell tumor – data of the MAHO/MAKEI-registry. Klin Padiatr 222:140–144

Haas RJ, Schmidt P, Gobel U et al (1995) Testicular germ cell tumors. Results of the GPO MAHO studies −82, −88, −92. Klin Padiatr 207:145–150

Hendricks M, Davidson A, Pillay K et al (2011) Carboplatin-based chemotherapy and surgery: a cost effective treatment strategy for malignant extracranial germ cell tumours in the developing world. Pediatr Blood Cancer 57:172–174

Horwich A, Oliver RTD, Wilkinson PM et al (2000) A Medical Research Council randomized trial of single agent carboplatin versus etoposide and cisplatin for advanced metastatic seminoma. Br J Cancer 83(12):1623–1629

Horwich A, Sleijfer DT, Fossa SD et al (1997) Randomized trial of bleomycin, etoposide, and cisplatin compared with bleomycin, etoposide, and carboplatin in good-prognosis metastatic nonseminomatous germ cell cancer: a multi-institutional Medical Research Council/European Organization for Research and Treatment of Cancer Trial. J Clin Oncol 15(5):1844–1852

Howlader N, Noone AM, Krapcho M et al (2011) SEER cancer statistics review, 1975–2008, National Cancer Institute. 2011, from http://seer.cancer.gov?csr/1975_2008/

Huddart SN, Mann JR, Gornall P et al (1990) The UK Children's Cancer Study Group: testicular malignant germ cell tumours 1979–1988. J Pediatr Surg 25(4):406–410

International Prognostic Factors Study Group (1997) International germ cell consensus classification: a prognostic factor-based staging system for metastatic germ cell cancers. J Clin Oncol 15:594–603

International Prognostic Factors Study Group (2010) Prognostic factors in patients with metastatic germ cell tumors who experienced treatment failure with cisplatin-based first-line chemotherapy. J Clin Oncol 28:4906–4911

Kaye SB, Mead GM, Fossa S et al (1998) Intensive induction-sequential chemotherapy with BOP/VIP-B compared with treatment with BEP/EP for poor-prognosis metastatic nonseminomatous germ cell tumor: a randomized Medical Research Council/European Organization for Research and Treatment of Cancer study. J Clin Oncol 16(2):692–701

Kollmannsberger C, Nichols C, Meisner C et al (2000) Identification of prognostic subgroups among patients with metastatic 'IGCCCG poor-prognosis' germ-cell cancer: an explorative analysis using cart modeling. Ann Oncol 11(9):1115–1120

Lewis CR, Fossa SD, Mead G et al (1991) BOP/VIP – a new platinum-intensive chemotherapy regimen for poor prognosis germ cell tumours. Ann Oncol 2(3):203–211

Li Y, Womer RB, Silber JH (2004) Predicting ototoxicity in children: influence of age and the cumulative dose. Eur J Cancer 40:2445–2451

Loehrer PJ Sr (1991) Etoposide therapy for testicular cancer. Cancer 67:220–224

Lopes LF, De Camargo B, Dondonis M et al (1995) Response to high-dose cisplatin and etoposide in advanced germ cell tumors in children: results of the Brazilian germ cell tumor study. Med Pediatr Oncol 25(5):396–399

Lopes LF, Macedo CRP, Pontes EM et al (2009) Cisplatin and etoposide in childhood germ cell tumor: Brazilian Pediatric Oncology Society Protocol GCT-91. J Clin Oncol 27:1297–1303

Lopes LF, Pontes EM, Macedo CRP (2010) Treatment with PE for intermediate-risk germ cell pediatric patients from the Brazilian GCT-99 protocol [abstract]. Pediatr Blood Cancer 55:805

Mann JR, Gray ES, Thornton C et al (2008) Mature and immature extracranial teratomas in children: the UK Children's Cancer Study Group experience. J Clin Oncol 26:3590–3597

Mann JR, Pearson D, Barrett A et al (1989) Results of the United Kingdom children's cancer study group's malignant germ cell tumor studies. Cancer 63:1657–1667

Mann JR, Raafat F, Robinson K et al (1998) UKCCSG's germ cell tumour (GCT) studies: improving outcome for children with malignant extracranial non-gonadal tumours – carboplatin, etoposide, and bleomycin are effective and less toxic than previous regimens. United kingdom Children's Cancer Study Group. Med Pediatr Oncol 30:217–227

Mann JR, Raafat F, Robinson K et al (2000) The United Kingdom Children's Cancer Study Group's Second Germ Cell Tumor Study: carboplatin, etoposide, and bleomycin are effective treatment for children with malignant extracranial germ cell tumors, with acceptable toxicity. J Clin Oncol 18:3809–3818

Marina N (2006) Prognostic factors in children with extragonadal malignant germ cell tumors: a Pediatric Intergroup study. J Clin Oncol 24:2544–2548

Marina N, Chang KW, Malogolowkin M et al (2005) Amifostine does not protect against the ototoxicity of high-dose cisplatin combined with etoposide and bleomycin in pediatric germ-cell tumors. Cancer 104:841–847

Marina N, Cushing B, Giller R et al (1999) Complete surgical excision is effective treatment for children with immature teratomas with or without malignant elements: a Pediatric Oncology Group/Children's Cancer Group Intergroup Study. J Clin Oncol 17:2137–2143

Mitchell PL, Al-Nasiri N, A'Hern R et al (1999) Treatment of nondysgerminomatous ovarian germ cell tumors: an analysis of 69 cases. Cancer 85(10):2232–2244

Motzer RJ, Nichols CJ, Margolin KA et al (2007) Phase III randomized trial of conventional-dose chemotherapy with or without high-dose chemotherapy and autologous hematopoietic stem-cell rescue as first-line treatment for patients with poor-prognosis metastatic germ cell tumors. J Clin Oncol 25:247–256

Nair R, Pai SK, Saikia TK et al (1994) Malignant germ cell tumors in childhood. J Surg Oncol 56(3):186–190

Neuwelt AE, Brock P (2010) Critical need for international consensus on ototoxicity assessment criteria. J Clin Oncol 28:1630–1632

Nichols CR, Williams SD, Loehrer PJ et al (1991) Randomized study of cisplatin dose intensity in poor-risk germ cell tumors: a Southeastern Cancer Study Group and Southwest Oncology Group protocol. J Clin Oncol 9(7):1163–1172

Norris HJ, Zirkin HJ, Benson WL (1976) Immature (malignant) teratoma of the ovary: a clinical and pathologic study of 58 cases. Cancer 37

Ozols RF, Ihde DC, Linehan WM et al (1988) A randomized trial of standard chemotherapy v a high-dose chemotherapy regimen in the treatment of poor prognosis nonseminomatous germ-cell tumors. J Clin Oncol 6(6):1031–1040

Palenzuela G, Martin E, Meunier A et al (2008) Comprehensive staging allows for excellent outcome in patients with localized malignant germ cell tumor of the ovary. Ann Surg 248:836–841

Patterson DM, Murugaesu N, Holden L et al (2008) A review of the close surveillance policy for stage I female germ cell tumors of the ovary and other sites. Int J Gynecol Cancer 18:43–50

Pinkerton CR, Broadbent V, Horwich A et al (1990) 'JEB' – a carboplatin based regimen for malignant germ cell tumours in children. Br J Cancer 62:257–262

Rexer FH (2005) International study on testicular cancer: EORTC 30983: randomized phase II/III study of Taxol-BEP versus BEP in patients with Intermediate Prognosis Germ Cell Cancer [German]. Der Urologe Ausg A 44(9):1064–1065

Rodriguez-Galindo C (2010) Clinical research in pediatric rare cancers [abstract]. Pediatr Blood Cancer 55:775–776

Rogers PC, Olson TA, Cullen JW et al (2004) Treatment of children and adolescents with stage II testicular and stages I and II ovarian malignant germ cell tumors: a Pediatric Intergroup Study – Pediatric Oncology Group 9048 and Children's Cancer Group 8891. J Clin Oncol 22:3563–3569

Russell WO, Cohen J, Enzinger F et al (1977) A clinical and pathological staging system for soft tissue sarcomas. Cancer 40:1562–1570

Samson MK, Rivkin SE, Jones SE (1984) Dose–response and dose-survival advantage for high versus low-dose cisplatin combined with vinblastine and bleomycin in disseminated testicular cancer. A Southwest Oncology Group study. Cancer 53(5):1029–1035

Saxman SB, Finch D, Gonin R et al (1998) Long-term follow-up of a phase III study of three versus four cycles of bleomycin, etoposide, and cisplatin in favorable-prognosis germ-cell tumors: the Indiana University experience. J Clin Oncol 16(2):702–706

Schlatter M, Rescorla F, Giller R et al (2003) Excellent outcomes in patients with stage I germ cell tumors of the testes: a study of the Children's Cancer Group/Pediatric Oncology Group. J Pediatr Surg 38

Schmidt P, Haas RJ, Gobel U et al (2002) Results of the German studies (MAHO) for treatment of testicular germ cell tumors in children – an update. Klin Padiatr 214:167–172

Sonneveld DJA, Hoekstra HJ, Van der Graaf WTA et al (2001) Improved long term survival of patients with metastatic nonseminomatous testicular germ cell carcinoma in relation to prognostic classification systems during the cisplatin era. Cancer 91(7):1304–1315

Tjulandin SA, Garin AM, Mescheryakov AA et al (1993) Cisplatin-etoposide and carboplatin-etoposide induction chemotherapy for good-risk patients with germ cell tumors. Ann Oncol 4:663–667

Travis LB, Beard C, Allan JM et al (2010) Testicular cancer survivorship: research strategies and recommendations. J Natl Cancer Inst 102:1114–1130

Williams SD, Birch R, Einhorn LH (1987) Treatment of disseminated germ-cell tumors with cisplatin, bleomycin, and either vinblastine or etoposide. N Engl J Med 316(23):1435–1440

Wood L, Kollmannsberger C, Jewett MA et al (2010) Canadian consensus guidelines for the management of testicular germ cell cancer. Can Urol Assoc J 4:E19–E38

Management of Disseminated Germ Cell Tumor in Adults

Craig R. Nichols, Christopher Porter, and Christian Kollmannsberger

Contents

C.R. Nichols, MD (✉)
Section of Hematology and Medical Oncology,
Virginia Mason Medical Center
and Testicular Cancer Commons,
Seattle, WA, USA
e-mail: craig.nichols@vmmc.org

C. Porter, MD
Section of Urology,
Virginia Mason Medical Center, Seattle, WA, USA
e-mail: christopher.porter@vmmc.org

C. Kollmannsberger, MD
Division of Systemic Therapy,
British Columbia Cancer Agency,
Vancouver, BC, Canada
e-mail: ckollmannsberger@bccancer.bc.ca

5.1 History of Clinical Trials in Disseminated Germ Cell Tumors

The modern era of management of germ cell tumors began with the discovery of cisplatin as a biologically active compound in the mid-1960s. Barnett Rosenberg, a biologist at Michigan State, was investigating the effect of electrical currents on bacteria. In his experiments, he used a platinum anode as he ran electrical current through plates of growing *E. coli*. He noticed around the platinum anode that the *E. coli* developed very elongated forms which on examination appeared as if the bacteria could not divide. He speculated that some elemental platinum was interfering with cell division. He isolated the cisplatin species. He further speculated that perhaps this compound could have an effective on rapidly dividing cells such as human cancer and went on to further demonstrate the effects of cisplatin on cultured tumor cells.

Subsequently cisplatin entered phase I cancer clinical trials. Within those trials, there were a few young men with disseminated germ cell tumors whom experienced marked responses. Coupled with these responses however were major dose-limiting side effects including vomiting and dehydration, nephrotoxicity, neurotoxicity, and ototoxicity.

It is important to recognize the backdrop of treatment in the 1960s and early 1970s. Systemic therapy was in its infancy with early single agents and combinations (actinomycin D, vinblastine, and bleomycin) demonstrating some efficacy with about half of testicular cancer patients experiencing partial or complete responses and perhaps 5–10 % of patients with disseminated disease actually experiencing long-term survival. The lack of an effective systemic therapy forced clinicians to apply radical regional therapies as active management of early stage disease and as potentially curative therapy for patients with known regional disease. In particular, extended radical retroperitoneal lymph node dissections were commonly applied to clinical stage I and II nonseminoma patients, and large field, high-dose radiation was commonly given for both early stage and regional seminomas.

In 1974 at Indiana University, Dr Lawrence Einhorn, a young medical oncologist, and Dr John Donohue, an experienced urologist, developed a multidisciplinary approach incorporating novel systemic combination chemotherapy including the newly identified active compound, cisplatin, along with adjunctive post-chemotherapy surgery for those with residual disease. The chemotherapy combination was four cycles of cisplatin, vinblastine, and bleomycin (PVB) plus maintenance therapy and was selected to incorporate active agents with nonoverlapping toxicity so that full doses of each constituent could be given.

The results were remarkable and changed forever the management of all stages of germ cell tumors. Of the original 47 patients, 33 (70 %) obtained a chemotherapy-induced complete remission (Einhorn and Donohue 1977). An additional 5 (11 %) patients obtained disease-free status after post-chemotherapy resection of residual disease. In long-term follow-up, 53 % of patients were in sustained remission representing a log higher cure rate than the contemporaneous dactinomycin-treated patients.

These results began to immediately inform the field both with reference to systemic disease and in terms of regional and adjuvant therapies. Urologists began to design and implement more limited RPLND templates that markedly reduced the uniform retrograde ejaculation seen with more extended operations. Radiation fields and doses were lessened, reducing short-term and presumably long-term toxicity. Clinical investigators were emboldened to consider just active surveillance for patients with clinical stage I disease knowing that there were highly effective salvage treatments available for the small portion of patients who relapsed.

5.2   History of Development of Systemic Therapy for Germ Cell Tumors

From the late 1970s through the mid-1980s, primary refinements to the basic PVB backbone were accomplished. Substantial toxicity reduction was accomplished in two studies from Indiana University. First, a randomized comparison of PVB with either 0.4 mg/kg of vinblastine or 0.3 mg/kg showed an improvement in toxicity for the 0.3 mg/kg arm but equivalent therapeutic effect. Second, the randomized comparison of induction therapy (4 cycles of PVB) with or without the original 2 years of maintenance vinblastine and bleomycin showed equivalent therapeutic effect of the 12-week program versus the 12-week plus maintenance program (Einhorn et al. 1981).

At the same time, etoposide was shown to have significant activity in recurrent germ cell tumors and in preliminary studies had a toxicity profile largely confined to hematopoietic toxicity. As such, etoposide was an ideal candidate for testing in the primary chemotherapy setting and to develop new salvage combinations. The seminal study reported by Dr. Stephen Williams compared four cycles of cisplatin and bleomycin and either vinblastine (PVB) or etoposide (BEP) as primary management of disseminated germ cell tumors. In this trial, 244 patients entered the trial. More patients became disease-free on BEP than PVB but not statistically significantly more. However, there were clinically and statistically meaningful reductions in myalgias, paresthesias, and abdominal cramps. BEP×4 became the standard therapy for bulky regional and disseminated germ cell tumors.

During this period, critically important non-chemotherapeutic developments profoundly shaped the field of systemic therapy for germ cell tumors. First, the important integration of expert post-chemotherapy resection of residual radiographic abnormalities was refined. Second, institutions began to look carefully at clinical parameters that could predict outcome with systemic therapy. Preliminary prognostic systems were developed at a number of institutions. The common feature was that it appeared distinct groups could be identified within the populations of patients with disseminated germ cell tumor; one larger group that predictably obtained disease-free status with systemic cisplatin-based chemotherapy and adjunctive surgery if required (>90 % cure rates) and a second smaller group of patients presenting with extensive anatomical disease and/or very high serum markers that did not reliably enter remission and had a substantial chance of dying of their illness (50–70 % cure rates). These efforts at defining prognosis ultimately culminated in a multinational effort that lead to the International Germ Cell Cancer Consensus Classification (IGCCCC) which aggregated clinical and follow-up data on more than 5,000 patients with seminoma and nonseminoma (IGCCCG 1997). The IGCCC project is summarized (Table 5.1) and identifies three risk categories: good risk, greater than 90 % longterm disease-free survival; intermediate risk, 70 % long-term survival; and poor risk, 50 % long-term disease-free survival.

With reliable, clinically based predictive systems, the clinical trials in adult germ cell tumors diverged. For the good-risk patients with a high predicted survival, the clinical question became "Can we retain this high cure rate and reduce short- and long-term toxicity?" For the poorerrisk categories wherein patients still had a significant chance of dying from the disease, the question became "Can we improve therapeutic outcomes through addition of new agents or intensifying dose and schedules of treatment?" With this approach, it was expected that toxicity of therapy would increase but that toxicity would hopefully be short term and balanced by higher cure rates.

For good-risk patients, comparative clinical trials accomplished some of these goals. Over the

Table 5.1 Prognostic-based staging system for metastatic germ cell cancer (International Germ Cell Cancer Collaborative Group)

Good-prognosis group	
Non-seminoma (56 % of cases)	*All of the following criteria:*
5-year PFS 89 %	Testis/retroperitoneal primary
5-year survival 92 %	No non-pulmonary visceral metastases
	AFP < 1,000 ng/mL
	hCG < 5,000 IU/L (1,000 ng/mL)
	LDH < 1.5 × ULN
Seminoma (90 % of cases)	*All of the following criteria:*
5-year PFS 82 %	Any primary site
5-year survival 86 %	No non-pulmonary visceral metastases
	Normal AFP
	Any hCG
	Any LDH
Intermediate prognosis group	
Non-seminoma (28 % of cases)	*All of the following criteria:*
5 years PFS 75 %	Testis/retroperitoneal primary
5-year survival 80 %	No non-pulmonary visceral metastases
	AFP > 1,000 and < 10,000 ng/mL or 10 Limited update March 2009
	hCG > 5,000 and < 50,000 IU/L or
	LDH > 1.5 and < 10 × ULN
Seminoma (10 % of cases)	*Any of the following criteria:*
5-year PFS 67 %	Any primary site
5-year survival 72 %	Non-pulmonary visceral metastases
	Normal AFP
	Any hCG
	Any LDH
Poor prognosis group	
Non-seminoma (16 % of cases)	*Any of the following criteria:*
5-year PFS 41 %	Mediastinal primary
5-year survival 48 %	Non-pulmonary visceral metastases
	AFP > 10,000 ng/mL or
	hCG > 50,000 IU/L (10,000 ng/mL) or
	LDH > 10 × ULN
Seminoma	
No patients classified as poor prognosis	

PFS progression-free survival, *AFP* alpha-fetoprotein, *hCG* human chorionic gonadotrophin, *LDH* lactate dehydrogenase years, randomized trials defined shorter chemo-therapy programs and a lower limit of therapy duration. The definitive trial came from Indiana University and the Southeastern Cancer Study Group where favorable patients were randomized to either BEP×4 versus BEP×3 (Einhorn et al. 1989). Preliminary results and now results obtained in long-term follow-up (Saxman et al. 1998) demonstrated that the arms were therapeu-tically equivalent but with obvious reduction of toxicity and costs related to eliminating the fourth cycle of BEP in favorable patients. Of note, how-ever, there was only one death related to chemo-therapy and no deaths in either arm from pulmonary complications. The next trial from Indiana and the SECSG attempted to continue the reduction of therapy in good-risk patients by randomly comparing patients receiving BEP×3 to three cycles of EP with the elimination of what may be the most troublesome drug in the combi-nation. The results were important. First, EP×3 was less effective in disease control than BEP×3 and no meaningful reduction of toxicity was seen (Culine et al. 2007; de Wit et al. 1997; Loehrer et al. 1995). Severe or worse non-hematologic toxicity was 9 % on EP×3 and 8 % on BEP. There was no significant difference in pulmonary toxicity and again no drug-related deaths related to pulmonary complications. There was a dou-bling of the number of patients who had unfavor-able outcomes (primary treatment failure, drug deaths, resected carcinoma relapse, and death) from 17 % with BEP×3 to 38 % in EP×3. The "floor" of minimal therapy for good-risk disease was established at BEP for three cycles.

A number of thoughtful clinical trials addressed different approaches to toxicity reduc-tion in this group of good-risk patients. The major trials are outlined in Table 5.2. Attempts to sub-stitute carboplatin for cisplatin (Bajorin et al. 1993; Bokemeyer et al. 1996; Horwich et al. 1997), attempts to eliminate bleomycin and attempts to reduce the etoposide dose, and attempts to shorten the 5-day schedule to 3 days were undertaken (de Wit et al. 2001). In sum-mary, all attempts either have failed to demon-strate therapeutic equivalence or significant toxicity advantages. The most recent attempt to improve the toxicity profile of BEP×3 while maintaining overall cure rates was the random-ized comparison of BEP×3 to EP×4. Culine and colleagues report this trial demonstrated a higher relapse and death rate with EP×4 and without a corresponding improvement in the toxicity pro-file relative to BEP×3. Since the early 1990s, three cycles of BEP has been the standard for patients with IGCCC good-risk disseminated germ cell tumors. In the USA, there have been no large randomized trials in good-risk disease since the comparison of BEP×3 to EP×3. Internationally, there are no trials underway and none planned.

For patients presenting with poor-risk catego-ries of disseminated germ cell tumors, clinical trials tested more intensive therapies in attempts to increase therapeutic effect. The results of major clinical trials are summarized in Table 5.3. Three US trials formed the basic pillars of inves-tigation in this era. The first intergroup trial com-pared standard therapy with BEP×4 to the combination of identical doses of etoposide and bleomycin but with a doubling of the cisplatin dose to 40 mg/m^2 daily×5 (Nichols et al. 1991). In this trial, the anticipated increased toxicity of the double-dose cisplatin was apparent with sig-nificant increases in ototoxicity, neurotoxicity, nephrotoxicity, and hematologic toxicity. However, there was no corresponding improve-ment in therapeutic outcome.

Second, the US intergroup mechanisms examined the role of ifosfamide in primary che-motherapy of poor-risk disseminated germ cell tumor. This trial again randomly compared standard therapy with BEP×4 to a new combi-nation of etoposide, ifosfamide, and cisplatin (VIP×4) (Nichols et al. 1998). Again, the antic-ipated increased toxicity was demonstrated in the VIP, particularly hematopoietic toxicity. However, again there was no difference in ther-apeutic effect with 64 % of patients being free of failure at 24 months on VIP compared to 60 % on BEP.

The final trial conducted through the inter-group mechanism was again comparing stan-dard BEP×4 to BEP×2 followed immediately by two cycles of high-dose chemotherapy with

Table 5.2 Poor-risk disease: selected randomized studies in patients with intermediate and poor prognosis nonseminoma

Author	Classification	Regime	Study objective	CR/PR rate %	Continuous CR/PR rate	Conclusion
Williams (1987)	Indiana All disseminated disease	PVB×4 BEP×4	Reduction of toxicity, retain efficacy	61 77 For high-volume pts		BEP×4 less toxic than PVB×4, BEP×4 trend towards improvement in advanced disease categories
Nichols (1991)	Indiana advanced	BEP×4 BEP2×4	Testing role of double-dose cisplatin	73 68	61 63 24 month f/u	No therapeutic differences, BEP2 much more toxic
Nichols (1998)	Indiana advanced	BEP×4 VIP×4	Role of ifosfamide-based therapy	60 64	60 63 60 month f/u	No significant therapeutic differences. VIP had more hematopoietic toxicity
Motzer (2007)	IGCCC Intermediate and poor risk	BEP×4 BEP×2+2×HDCT	Role of high-dose chemotherapy	56 55 CR rate	57 60 51 month f/u	No overall therapeutic difference, BEP×2+2×HDCT more toxic
Culine (2008)	IGCCCG Intermediate and poor risk	BEP×4 CISCA/VB	Role of dose-dense multiagent chemotherapy	65 57	69 59 5 years overall survival	CISCA/VB more toxic, BEP×4 trends towards more effective
Kaye (1998)	EORTC Poor risk	BEP/EP×4 BOP/VIP-B	Role of dose-dense multiagent chemotherapy	57 54	60 53 1 year failure-free survival	BOP/VIP-B more toxic, failed to improve outcomes relative to BEP/EP×4
de Wit (2012)	IGCCC Intermediate risk only	BEP×4 T-BEP×4	Role of taxol-based therapy in intermediate-risk disease	70 60	71 79 3 years progression-free survival	T-BEP required growth factors. No statistical improvement in therapeutic outcomes

carboplatin, etoposide, and cyclophosphamide with stem cell transplant (Motzer et al. 2007). Toxicity was increased with the incorporation of high-dose chemotherapy, but there was not an excess of therapy of related deaths on the high-dose arm. Unfortunately, despite increasing chemotherapy intensity to levels necessitating stem cell support, no consistent therapeutic effect could be discerned. With a median follow-up of 51 months, 69 % of patients were alive on BEP×4 and 68 % of patients receiving BEP×2 and two high-dose chemotherapy cycles.

A series of non-US trials have approached this group of patients with decreased overall survival as well. Attempts to add active agents and increase dose density and dose intensity all have added toxicity and cost without improving overall survival. Of particular note is the recent EORTC trial adding paclitaxel and filgrastim to standard BEP in patients with IGCCCC intermediate-risk germ cell tumors (de Wit et al. 2012). Despite over 10 years of dogged effort, the trial failed to accrue sufficient patients to be definitive. Toxicity and cost were higher in the

Table 5.3 Good-risk disease: selected randomized studies in patients with good prognosis nonseminoma

Author	Classification	Regime	Study objective	CR/PR rate %	Continuous CR/PR rate	Conclusion
Einhorn et al. (1989)	Indiana	PEB×4	Reduction of number of cycles	97	88	BEP×3 equally effective to BEP×4
Saxman et al. (1998)		PEB×3		98	87	
Bosl et al. (1988)	MSKCC	PE×4	Testing of a new 2-drug combination	93	82	PE×4 equally effective to 3×VAB-6×3
		VAB-6×3		96	85	
de Wit et al. (1997)	EORTC	PE$_{360}$B×4	Evasion of bleomycin	95	91	PE×4 inferior to BEP×4
		PE$_{360}$×4		87	83	
Loehrer et al. (1995)	Indiana	PEB×3	Evasion of bleomycin	94	86	PE×3 inferior to 3×BEP
		PE×3		88	69	
de Wit et al. (2001)	IGCCCG	PEB×3+1PE	Reduction of number of cycles	73	91	PEB×3 equally effective to BEP×4
		PEB×3		71	89	
Culine et al. (2003)	IGCCCG	PEB×3	Evasion of bleomycin	92	90	Similar high response rates, study too small to comment on relapse-free or overall survival
		PE×4		91	84	
Bajorin et al. (1993)	MSKCC	PE×4	Carboplatin versus cisplatin	88	87	CE×4 inferior to PE×4
		CE×4		80	76	
Horwich et al. (1997)	MRC/EORTC	PEB×4	Carboplatin versus cisplatin	94	91	CEB×4 inferior to BEP×4
		CEB×4		87	77	
Bokemeyer et al. (1996)	Indiana	PEB×3	Carboplatin versus cisplatin	97	86	CEB inferior to BEP
		CEB×4		96	68	

T-BEP arm, and there was not a demonstrable improvement in any of the predefined therapeutic endpoints. Since 1987 with the original trial of Williams and colleagues, BEP×4 has remained the standard of chemotherapy management for patients with IGCCCC intermediate- and poor-risk patients.

Despite lack of demonstrable progress in clinical trials, there appears to be continuous improvements in outcomes. Certainly in single institutions or in populations with centralized management of germ cell tumors, there has been substantial improvement in outcomes relative to the IGCCCCG estimates. Long-term survival in patients with regional and good-risk disseminated nonseminoma now routinely exceeds 95 %. Several large regional consortiums are reporting overall survival in intermediate- and poor-risk nonseminoma approaching 90 and 70 %, respectively. Large population-based reports in seminoma suggest that death from any stage seminoma is extremely rare and cure rates in regional or disseminated seminoma treated with standard

chemotherapy exceed 95 %. IGCCCCG estimates of 90, 70, and 50 % for good-, intermediate- and poor-risk patients were originally reported in the mid-1990s and were based on clinical data from approximately 10 years earlier. The cause of these seeming improvements likely is related to favorable stage migration, improved salvage therapy, more consistent application of adjunctive surgery, and consistent dosing and delivery of BEP (Kollmannsberger et al. 2010).

5.3 Initial Evaluation/ Survivorship Planning

In caring for patients with germ cell tumors, providers often are overly preoccupied with the explosive behavior occasionally seen in such patients. Such biological behavior almost always is associated with obvious widely disseminated disease which can be a true oncologic emergency. However, the vast majority of patients with regional or disseminated disease have disease that is reasonably paced and have the associated luxury of being able to briskly, but thoroughly evaluate the patient, gather other opinions or consults and plan both appropriate treatment and high-quality survivorship.

The primary goal of evaluation of the patient with regional or disseminated disease is to formulate an appropriate chemotherapy treatment plan that maximizes cure rates and leads to high-quality survival (Krege et al. 2008a, b; Wood et al. 2010). Formal stage and International Germ Cell Cancer Consensus Conference (IGCCCC) risk assessment at this time informs the development of the treatment plan. These are several primary questions to be answered at the time of initial suspicion of regional or disseminated testicular cancer. The evaluation required at this point is not complicated. The evaluation depends almost entirely on history (including survivorship profile) and physical exam (including examination of the remaining testis and clinical assessment of pulmonary reserve), simple laboratory parameters to determine liver and renal function, anatomical imaging and serum HCG, AFP, and LDH. It should be emphasized that the determination of IGCCCC risk relative to marker elevation refers to the stable post-orchiectomy markers. Anatomical imaging consists of chest, abdomen, and pelvic CT with intravenous and oral contrast. The routine use of positron emission tomography (PET) increases radiation exposure, does not provide additional meaningful clinical information, and is not recommended. Routine imaging of brain or bone is not recommended unless there are clinical concerns regarding these sites or there are widely disseminated, high-volume metastases (IGCCCC poor-risk disease) (Table 5.1).

Patients occasionally present with signs and symptoms of overwhelming metastatic disease and have not had an orchiectomy for tissue diagnosis. In this setting, orchiectomy is not mandatory in the pre-chemotherapy treatment period if there is a diagnostic elevation of tumor markers (HCG, AFP) or a biopsy of a metastatic deposit has established the diagnosis. The testicular primary is then removed after completion of chemotherapy often in conjunction with the post-chemotherapy RPLND.

During the initial evaluation of regional and disseminated disease, one must consider the physiological suitability of the patient for receipt of multiagent chemotherapy particularly cisplatin and bleomycin. This evaluation is critical to treatment planning. Details of evaluation for such suitability are outlined in the section on practical delivery of BEP.

5.4 Formulating a Treatment and Survivorship Plan

The evaluation as outlined will provide the substrate to determine the survivorship and treatment plan. Key issues that must be addressed initially are fertility and fertility preservation, physiological, financial, psychological, and social barriers to effective treatment, patient education, and patient engagement. For patients without completed families and who desire this option, guidelines and experts recommend consideration of semen analysis and cryopreservation prior to definitive treatment (Krege et al. 2008a, b; Wood et al. 2010). It should be emphasized that, in the USA, insurance coverage for semen cryopreservation is uneven, and the cost is often borne by the individual. Several organizations have worked to provide free or reduced cost access to fertility services (LIVE*STRONG* Fertile Hope and Oncofertility). Most testicular cancer patients do recover endogenous sperm production to sufficient levels after chemotherapy or radiation, and some of those who require post-chemotherapy surgery are able to have nerve-sparing procedures that allow for antegrade ejaculation.

5.5 Treatment Recommendations for Nonseminoma by Stage

5.5.1 Practical Delivery of BEP

Best chemotherapy outcomes in regional and disseminated germ cell tumors depend on timely and safe delivery of standard cisplatin, etoposide, and bleomycin. Doses and schedules of BEP are standardized. There are clinical data suggesting that significant deviations in dose and timing from the well-defined schedule have a meaningful negative impact on outcomes. As well, standard dose and schedule BEP when given by attentive and experienced chemotherapists with modern supportive care is remarkably safe with low incidence of intolerable nausea and vomiting and neutropenic fever and less than 1 % incidence of fatal complications of treatment.

5.5.1.1 When to Seek Consultation with High-Volume Center

For many years, led by Dr. Lawrence Einhorn and Dr. Stephen Williams at Indiana University, the international academic testicular cancer community has been extremely open to solicitation of opinions from the community or direct referral of any patient for a second opinion or application of specialized expertise. Our recommendations are to take advantage of these services as often as is needed. In particularly, we would recommend referral of all patients requiring tertiary services (clinical trials, post-chemotherapy assessment and surgery, patients with suspected or proven recurrent disease needing salvage treatments and patients with complex, high-risk clinical scenarios including all patients with initial intermediate- and poor-risk disease in particular those with high HCG, patients presenting with high-volume abdominal disease, patients with significant comorbid conditions, and patients with late recurring germ cell tumors).

5.5.1.2 Practical Management of Hematologic Complications

In good-risk patients receiving standard BEP, the primary hematologic complication is transient neutropenia. Significant anemia or thrombocytopenia is rare. Neutropenic fever is uncommon and occurs in about 15 % of cycles in good-risk patients. In patients with more advanced disease or associated poor performance status, the incidence can approach 30–40 %.

Most centers do not use prophylactic hematopoietic growth factors in patients with good-risk disease and reserve hematopoietic growth factors for those patients presenting with intermediate- or poor-risk disease, those rare patients who present with significantly impaired performance status, those receiving salvage chemotherapy, and those with a history of abdominal/pelvic radiation. Those good-risk patients who experience neutropenic fever with initial chemotherapy cycles receive hematopoietic growth factors as secondary prophylaxis.

Initiation of chemotherapy should not be delayed based on arbitrary neutrophil recovery. Cycles should start every 21 days without dose attenuation or delay independent of neutrophil count. Holidays or vacations are not suitable reasons to alter the schedule. Nursing colleagues should receive specific instructions regarding treating patients with sometimes astonishing low counts and be reassured that these young patients typically have very rapid hematopoietic recovery without initiation of growth factors or dose attenuation or delay.

5.5.1.3 Management of Chemotherapy-Induced Nausea and Vomiting

BEP combination chemotherapy is highly emetogenic and requires appropriate levels of antiemetic support. As well, delayed nausea and vomiting can be pronounced in this population receiving multiday cisplatin. Optimizing control of both acute and delayed nausea and vomiting is critical to managing fluid balance and keeping patients on schedule. A typical BEP order set should include appropriate antiemetics which includes 5HT-3 antagonists, steroids, and, in many institutions, aprepitant.

5.5.1.4 Management of Potential Nephrotoxicity

Aggressive pre- and post-chemotherapy hydration is the key to minimizing nephrotoxic consequences of cisplatin-based chemotherapy. If the patient is receiving cisplatin-based chemotherapy as an outpatient, at least one liter of normal saline should be given prior to initiation of cisplatin and 1.5 to two additional liters after cisplatin each day. Patients should be encouraged to drink liquids aggressively and report significant vomiting immediately. Patients frequently gain a significant amount of "water" weight over the course of each 5-day session, but this mobilizes rapidly and does not require diuretics for management.

Invariably, patients will have some wasting of magnesium, calcium, and potassium over the course of treatment. Rarely is this of clinical consequence. The common practice of adding magnesium to IV fluids does not prevent ongoing renal losses and is not recommended. Mannitol to promote diuresis has been associated in a small randomized trial with increased renal complications (Morgan et al. 2012; Leu and Baribeault 2010; Santoso et al. 2003). Most centers with significant experience do not add mannitol to the IV fluids used in hydration.

5.5.1.5 Management of Potential Bleomycin Complications

Bleomycin is an essential drug for optimal management of testicular cancer and can be given safely in almost all patients. Anaphylaxis is very rare and can occur at any dose. Thus, the older practice of bleomycin test doses or skin tests are not used in most centers. In published experiences of three cycles of BEP, deaths from toxicity of any type are very rare, and in particular it is difficult to find patients in a modern series who have died of bleomycin toxicity. In large experiences using four cycles of BEP, less than 1 % of patients are reported to have died from bleomycin pulmonary toxicity. That said, there are some common precautions and ongoing assessments that can help identify early those few patients who may be developing preliminary signs of toxicity. The first challenge is to identify patients who may be at heightened risk for bleomycin toxicity in particular patients with demonstrable severe lung compromise, patients with significantly reduced renal function, and older patients. In addition, in good-risk patients, one must balance the risk of bleomycin (BEP × 3) against the toxicity of the obligate additional course of cisplatin (EP × 4) if one were to delete bleomycin as well as the potential of losing therapeutic punch by eliminating bleomycin from the regimen. Second, it must be recognized that managing potential bleomycin pulmonary toxicity is primarily a clinical challenge and it is very difficult

to predict bleomycin lung disease based primarily on changes in pulmonary function tests. A practical approach to this problem is to reserve EP or VIP for those few patients who have significant contraindications to bleomycin. All patients who receive bleomycin should be informed of the necessity of reporting pulmonary symptoms such as a dry cough. Patients should be queried for such symptoms on a regular basis during chemotherapy. Patients should also have a pulmonary examination by the provider with each cycle of therapy to discern if there is an inspiratory lag or the development of dry or "Velcro" rales, and if these symptoms and sign appear, bleomycin should be discontinued. Common practice for patients scheduled to receive four cycles of therapy is to perform pulmonary function tests at the initiation of therapy and the beginning of the fourth cycle. If there has been a significant decline in the diffusion capacity (>30–40 %), bleomycin is held. Also, for intermediate- and poor-risk patients receiving four cycles of BEP in whom a significant post-chemotherapy surgery is anticipated, common practice has been to hold the eleventh and twelfth doses of bleomycin.

Bleomycin skin changes are common and usually transitory. Bleomycin should not be held or eliminated based on skin streaking, painful nodules on digits, or discoloration.

5.6 Post-chemotherapy Management

At the completion of chemotherapy for regional or disseminated germ cell tumors, a careful reassessment is required within 4 weeks of completion of chemotherapy to establish formally the post-chemotherapy status and the possible need for further adjunctive treatments and to establish a plan for follow-up and long-term survivorship.

Nonseminoma patients completing chemotherapy for nonseminoma must be evaluated for residual disease and the need for post-chemotherapy surgery. Residual radiographic abnormalities in patients with nonseminoma are not necessarily biologically inert. Approximately

40–45 % of patients have residual chemotherapy-insensitive teratoma and 5–10 % have residual viable germ cell cancer. The remainder have necrotic fibrotic residual without biologically active elements. There is universal consensus among experts and across guidelines that patients with residual radiographic findings greater than one cm on axial imaging have full resection of residual radiographic abnormalities (usually RPLND). For patients obtaining a complete radiographic remission (residual less than one cm), there is some controversy regarding the necessity of resection. Two large retrospective experiences inform this debate. These large series demonstrate that the abdominal recurrence rate is less than 10 % for patients observed after obtaining complete chemotherapy-induced remission, and there does not appear to be an enhanced risk of late relapse or death in patients so managed. There are no randomized comparisons of selective post-chemotherapy observation versus surgery for all patients after chemotherapy. Opinions and institutional guidelines vary markedly (Krege et al. 2008a, b; Wood et al. 2010). Most guidelines and most high-volume centers recommend selective application of surgery in the post-chemotherapy setting reserving this for patients with greater than one cm residual disease on axial imaging. There are some institutions that consider initial IGCCC risk as a factor in recommending surgery and often offer surgery in patients with initial intermediate- or poor-risk disease despite achieving complete remission with primary chemotherapy.

Seminoma patients completing chemotherapy for seminoma also must be evaluated to establish the post-chemotherapy status and need for additional evaluations. In seminoma, most centers and most guidelines use a cutoff of 3 cm disease on axial imaging to gauge the need for further evaluations. For patient with less than 3 cm residual, risk of recurrence is very low and further evaluations of aggressive posttreatment imaging is not recommended. In patients with greater than 3 cm residual abnormalities, there is a higher chance of harboring residual seminoma and further evaluation or close follow-up is recommended. Currently, where positive emission tomography (PET) is available, it is recommended that a PET scan be obtained at the 6–8 week post-chemotherapy point. If negative, the patient should be observed as a low-risk seminoma patient. For those with positive or equivocal PET, it is recommended that the patient undergo biopsy or resection to clarify pathologically the nature of the residual mass or undergo serial anatomic imaging to establish a pattern of growth. There are a number of false-positive PET determinations even at experienced centers, and salvage therapy should not be initiated without evidence of residual seminoma or evidence of serial growth of the disease.

Follow-Up Schedules and Imaging. There is no high-grade evidence to guide management of patients after completion of primary therapy for disseminated disease in terms of follow-up schedules and imaging. Nearly all of this has arisen from individual institution's experience and biases. Published schedules range from low intensity follow-up and imaging over a 3–5-year period to intense schedules including more than 20 CTs over a 10-year period. The single randomized comparison available in this clinical arena is reported by Rustin and colleagues (Rustin et al. 2007). In CSI nonseminoma, imaging schedules of two abdominal CTs were compared to five abdominal CTs. No difference was identified in the proportion of patients relapsing with intermediate- or poor-risk disease or a difference in the size of the abdominal relapse. Appropriate concerns about cost, false-positives, and exposure to ionizing radiation are being raised. Based on contemporary discussions with international experts and known periods of risk, our groups recommend a parsimonious approach with most imaging confined to primary periods of risk of recurrence (years 1 and 2), selective imaging of areas at highest risk (abdominal CTs without chest or pelvic CTs), and exploration of technologies that reduce exposure to ionizing radiation such as low-dose CT and MRI. Beyond year 2, the focus largely becomes management of the often subtle effects of treatment of the disease itself (hypogonadism, infertility, metabolic syndrome, social emotional consequences) often in conjunction with the primary care team.

References

Bajorin DF, Sarosdy MF, Pfister DG et al (1993) Randomized trial of etoposide and cisplatin versus etoposide and carboplatin in patients with good-risk germ cell tumors: a multiinstitutional study. J Clin Oncol 11(4):598–606

Bokemeyer C, Köhrmann O, Tischler J et al (1996) A randomized trial of cisplatin, etoposide and bleomycin (PEB) versus carboplatin, etoposide and bleomycin (CEB) for patients with 'good-risk' metastatic nonseminomatous germ cell tumors. Ann Oncol 7(10):1015–1021

Bosl GJ, Geller NL, Bajorin D et al (1988) A randomized trial of etoposide + cisplatin versus vinblastine + bleomycin + cisplatin + cyclophosphamide + dactinomycin in patients with good-prognosis germ cell tumors. J Clin Oncol 6(8):1231–1238

Culine S, Lortholary A, Voigt JJ et al (2003) Cisplatin in combination with either gemcitabine or irinotecan in carcinomas of unknown primary site: results of a randomized phase II study–trial for the French Study Group on Carcinomas of Unknown Primary (GEFCAPI 01). J Clin Oncol 21(18):3479–3482

Culine S, Kerbrat P, Kramar A et al (2007) Refining the optimal chemotherapy regimen for good-risk metastatic nonseminomatous germ-cell tumors: a randomized trial of the Genito-Urinary Group of the French Federation of Cancer Centers (GETUG T93BP). Ann Oncol 18(5):917–924

Culine S, Kramar A, Théodore C et al (2008) Randomized trial comparing bleomycin/etoposide/cisplatin with alternating cisplatin/cyclophosphamide/doxorubicin and vinblastine/bleomycin regimens of chemotherapy for patients with intermediate- and poor-risk metastatic nonseminomatous germ cell tumors: Genito-Urinary Group of the French Federation of Cancer Centers Trial T93MP. J Clin Oncol 26(3):421–427

de Wit R, Stoter G, Kaye SB et al (1997) Importance of bleomycin in combination chemotherapy for good-prognosis testicular nonseminoma: a randomized study of the European Organization for Research and Treatment of Cancer Genitourinary Tract Cancer Cooperative Group. J Clin Oncol 15(5): 1837–1843

de Wit R, Roberts JT, Wilkinson PM et al (2001) Equivalence of three or four cycles of bleomycin, etoposide, and cisplatin chemotherapy and of a 3- or 5-day schedule in good-prognosis germ cell cancer: a randomized study of the European Organization for Research and Treatment of Cancer Genitourinary Tract Cancer Cooperative Group and the Medical Research Council. J Clin Oncol 19(6):1629–1640

de Wit R, Skoneczna I, Daugaard G et al (2012) Randomized phase III study comparing paclitaxel-bleomycin, etoposide, and cisplatin (BEP) to standard BEP in intermediate-prognosis germ-cell cancer: intergroup study EORTC 30983. J Clin Oncol 30(8): 792–799

Einhorn LH, Donohue J (1977) Cis-diamminedichloro-platinum, vinblastine, and bleomycin combination chemotherapy in disseminated testicular cancer. Ann Intern Med 87(3):293–298

Einhorn LH, Williams SD, Troner M et al (1981) The role of maintenance therapy in disseminated testicular cancer. N Engl J Med 305(13):727–731

Einhorn LH, Williams SD, Loehrer PJ et al (1989) Evaluation of optimal duration of chemotherapy in favorable-prognosis disseminated germ cell tumors: a Southeastern Cancer Study Group protocol. J Clin Oncol 7(3):387–391

Horwich A, Sleijfer DT, Fosså SD et al (1997) Randomized trial of bleomycin, etoposide, and cisplatin compared with bleomycin, etoposide, and carboplatin in good-prognosis metastatic nonseminomatous germ cell cancer: a Multiinstitutional Medical Research Council/European Organization for Research and Treatment of Cancer Trial. J Clin Oncol 15(5):1844–1852

International Germ Cell Cancer Collaborative Group (1997) International Germ Cell Consensus Classification: a prognostic factor-based staging system for metastatic germ cell cancers. J Clin Oncol 15(2):594–603

Kaye SB, Mead GM, Fossa S et al (1998) Intensive induction-sequential chemotherapy with BOP/VIP-B compared with treatment with BEP/EP for poor-prognosis metastatic nonseminomatous germ cell tumor: a Randomized Medical Research Council/European Organization for Research and Treatment of Cancer study. J Clin Oncol 16(2):692–701

Kollmannsberger C, Daneshmand S, So A et al (2010) Management of disseminated nonseminomatous germ cell tumors with risk-based chemotherapy followed by response-guided postchemotherapy surgery. J Clin Oncol 28(4):537–542

Krege S, Beyer J, Souchon R et al (2008a) European consensus conference on diagnosis and treatment of germ cell cancer: a report of the second meeting of the European Germ Cell Cancer Consensus group (EGCCCG): part I. Eur Urol 53(3):478–496

Krege S, Beyer J, Souchon R et al (2008b) European consensus conference on diagnosis and treatment of germ cell cancer: a report of the second meeting of the European Germ Cell Cancer Consensus group (EGCCCG): part II. Eur Urol 53(3):497–513

Leu L, Baribeault D (2010) A comparison of the rates of cisplatin (cDDP)–induced nephrotoxicity associated with sodium loading or sodium loading with forced diuresis as a preventative measure. J Oncol Pharm Pract 16(3):167–171

Loehrer PJ Sr, Johnson D, Elson P et al (1995) Importance of bleomycin in favorable-prognosis disseminated germ cell tumors: an Eastern Cooperative Oncology Group trial. J Clin Oncol 13(2):470–476

Morgan KP, Buie LW, Savage SW (2012) The role of mannitol as a nephroprotectant in patients receiving cisplatin therapy. Ann Pharmacother 46(2): 276–281

Motzer RJ, Nichols CJ, Margolin KA et al (2007) Phase III randomized trial of conventional-dose chemotherapy with or without high-dose chemotherapy and autologous hematopoietic stem-cell rescue as first-line treatment for patients with poor-prognosis metastatic germ cell tumors. J Clin Oncol 25(3):247–256

Nichols CR, Williams SD, Loehrer PJ et al (1991) Randomized study of cisplatin dose intensity in poor-risk germ cell tumors: a Southeastern Cancer Study Group and Southwest Oncology Group protocol. J Clin Oncol 9(7):1163–1172

Nichols CR, Catalano PJ, Crawford ED et al (1998) Randomized comparison of cisplatin and etoposide and either bleomycin or ifosfamide in treatment of advanced disseminated germ cell tumors: an Eastern Cooperative Oncology Group, Southwest Oncology Group, and Cancer and Leukemia Group B Study. J Clin Oncol 16(4):1287–1293

Rustin GJ, Mead GM, Stenning SP et al (2007) Randomized trial of two or five computed tomography scans in the surveillance of patients with stage I nonseminomatous germ cell tumors of the testis: Medical Research Council Trial TE08, ISRCTN56475197 – the National Cancer Research Institute Testis Cancer Clinical Studies Grou. J Clin Oncol 25(11):1310–1315

Santoso JT, Lucci JA 3rd, Coleman RL et al (2003) Saline, mannitol, and furosemide hydration in acute cisplatin nephrotoxicity: a randomized trial. Cancer Chemother Pharmacol 52(1):13–18

Saxman SB, Finch D, Gonin R, Einhorn LH (1998) Long-term follow-up of a phase III study of three versus four cycles of bleomycin, etoposide, and cisplatin in favorable-prognosis germ-cell tumors: the Indian University experience. J Clin Oncol 16(2):702–706

Williams SD, Birch R, Einhorn LH et al (1987) Treatment of disseminated germ-cell tumors with Cisplatin, Bleomycin, and either Vinblastine or Etoposide. N Engl J Med 316:1435–1440

Wood L, Kollmannsberger C, Jewett M et al (2010) Canadian consensus guidelines for the management of testicular germ cell cancer. Can Urol Assoc J 4(2): e19–e38

Clinical Treatment of Adult Women with Ovarian Germ Cell Tumors

6

David M. Gershenson and Jubilee Brown

Contents

6.1 Pathology

It is important to distinguish the primitive germ cell tumors (dysgerminoma, yolk sac tumor, embryonal carcinoma, polyembryoma, nongestational choriocarcinoma, and mixed germ cell tumor) and biphasic or triphasic teratomas (immature teratoma) from the even less common monodermal teratoma and somatic-type tumor category, which includes such tumor types as carcinoid, malignancies arising in a benign cystic teratoma (melanoma, carcinoma, sarcoma, etc.), or primitive neuroectodermal tumor (Tavassoli and Deville 2003; Roth and Talerman 2006). This chapter will discuss only the clinical management of the first two types.

6.2 Prognostic Factors

Although prognostic factors for testicular cancer have been identified relatively easily, doing so for ovarian germ cell tumors has been considerably more challenging related to the rarity of this entity. However, over the past few years, a few large-scale studies have identified key factors, including International Federation of Gynecology and Obstetrics staging system's (FIGO) stage, lymph node involvement, histologic type, residual disease, type of chemotherapy, and elevation of serum tumor markers (Lai et al. 2005; Murugaesu et al. 2006; Kumar et al. 2008). Lai et al. reported that advanced FIGO stage and

D.M. Gershenson, MD (✉) • J. Brown, MD
Department of Gynecologic Oncology and
Reproductive Medicine, The University of Texas MD
Anderson Cancer Center, Houston, TX, USA
e-mail: dgershen@mdanderson.org;
jbbrown@mdanderson.org

A.L. Frazier, J.F. Amatruda (eds.), *Pediatric Germ Cell Tumors*, Pediatric Oncology 1,
DOI 10.1007/978-3-642-38971-9_6, © Springer-Verlag Berlin Heidelberg 2014

nondysgerminoma/immature teratoma histology were associated with a significantly increased risk of treatment failure. Murugaesu et al. found that FIGO stage and combined elevations of hCG and AFP were prognostic. Kumar et al. recently reported that, after controlling for age, race, FIGO stage, histology, and grade, the presence of lymph node involvement is an independent predictor of poor survival, with a hazard ratio of 2.87 (Kumar et al. 2008). In a study of 1984 male patients with metastatic germ cell tumors of both gonadal and extragonadal sites, The International Prognostic Factors Study Group found that histology, primary tumor location, response, progression-free interval after first-line treatment, levels of alpha fetoprotein or human chorionic gonadotropin, and the presence of liver, bone, or brain metastases at salvage were identified as independent prognostic factors and were used to establish a prognostic model (The International Prognostic Factors Study Group 2010).

6.3 Primary Surgical Management

There is no standard evaluation for a young woman suspected of having a malignant ovarian germ cell tumor. Generally, the signs and symptoms are indistinguishable from those associated with other ovarian tumor types. Because ovarian germ cell tumors are, on average, larger than the more common epithelial ovarian carcinomas—a mean of 16 cm in diameter in the M.D. Anderson series—pelvic/abdominal pain may be more prominent in this patient cohort. In addition, fever or abnormal vaginal bleeding may occur in a small proportion of patients.

In our view, a minimum work-up for a young patient with an adnexal mass suspicious for malignancy should include routine blood studies and chest x-ray, serum tumor markers, transvaginal sonography (to elucidate the specific characteristics of the ovarian mass(es)), and computed tomography of the abdomen/pelvis to provide information on potential upper abdominal or retroperitoneal involvement.

Primary surgery serves several purposes, including the resection of tissue to make an accurate histological diagnosis, the prospect of comprehensive surgical staging in early stage disease, and the opportunity for cytoreductive surgery in the event that advanced stage disease is encountered. Regardless of stage, most young women who are found to have a malignant ovarian germ cell tumor may be treated with fertility-sparing surgery given the characteristics of this tumor type. Typically the surgeon relies on the intraoperative pathology report provided during surgery. Though it is sometimes difficult for the general pathologist to characterize the histology based solely on the microscopic intraoperative evaluation, this is often the only information the surgeon has upon which to make decisions regarding fertility preservation and the extent of surgery, as the diagnosis cannot be made based on gross inspection alone.

Contemporary surgical principles are primarily based on information derived from early case series on the operative and pathological findings in patients with malignant ovarian germ cell tumors (Norris et al 1976; Kurman and Norris 1976a, b, c; Kurman and Norris 1977; Jimerson and Woodruff 1977a, b; Gershenson et al. 1983, 1984, 1986a, b). These reports collectively indicate that malignant ovarian germ cell tumors are usually quite large, are frequently confined to the ovary (stage I), and are unilateral. The exception to the latter is in the case of pure dysgerminoma or mixed germ cell tumor with a dysgerminoma component, in which approximately 10–15 % of cases are associated with bilateral ovarian involvement. In addition, an ipsilateral or contralateral benign cystic teratoma may be present in 5–10 % of cases.

6.3.1 Fertility-Sparing Surgery

Most patients with malignant ovarian germ cell tumors are candidates for fertility-sparing surgery related to the combined factors of young age, unilateral ovarian involvement, and preponderance of stage I disease. In their report of 182 patients with tumors grossly confined to the ovary, Kurman and Norris observed no negative influence on prognosis associated with fertility-sparing surgery (Kurman and Norris 1977). For the past few decades, fertility-sparing surgery consisting of unilateral salpingo-oophorectomy has become the standard management for young adult women who wish to preserve fertility. Even for patients with advanced stage disease, it is frequently possible to preserve the uterus and normal contralateral ovary. Several series report the rate of fertility-sparing surgery of 55–81 % of patients (Zanetta et al. 2001; Gershenson et al. 2007; Weinberg et al. 2011). These studies do include patients treated as remotely as the 1970s, when fertility preservation was not a consideration, so current percentages of successful fertility preservation may be higher. In a small proportion of these patients, both ovaries were removed because of tumor involvement or disseminated disease. In select cases, even ovarian cystectomy may be adequate surgical treatment of the involved ovary (Beiner et al. 2004). The presence of dysgenetic gonads, if diagnosed preoperatively or at surgery, however, would preclude such management.

6.3.2 Comprehensive Surgical Staging

The overarching principle of comprehensive surgical staging for women with apparent early stage malignant ovarian germ cell tumors has not been well studied in this tumor type but rather is based on experience with the more common epithelial ovarian cancers. Thus, most gynecologic oncologists believe that such management most accurately determines the extent of disease, provides prognostic information, and potentially guides postoperative management. A critical component of this practice is the availability and performance of frozen section examination intraoperatively by an expert gynecologic pathologist. The standard approach consists of peritoneal cytologic washings (or submission of ascites fluid), systematic inspection and palpation of all peritoneal surfaces, multiple biopsies of pelvic and abdominal peritoneal surfaces, omentectomy, and retroperitoneal lymphadenectomy,

including bilateral pelvic and para-aortic lymph node regions.

A very different perspective appears to prevail in the pediatric surgical community. Based on an intergroup study in which deviations from standard surgical guidelines as detailed above did not adversely influence survival, Billmire et al. proposed a different set of surgical guidelines: collection of ascites or cytologic washings, examination of the peritoneal surfaces with biopsy or excision of any nodules (i.e., no random biopsies of normal-appearing peritoneum), examination and palpation of retroperitoneal lymph nodes and sampling of any firm or enlarged nodes (i.e., no systematic lymphadenectomy), inspection and palpation of omentum with removal of any adherent or abnormal areas, biopsy of any abnormal areas, and complete resection of the tumor-containing ovary with sparing of the fallopian tube if not involved (Billmire et al. 2004). It does not make sense that the surgical management should differ for children versus adults; thus, further study of the role of comprehensive surgical staging is warranted. In the meantime, most adult patients who undergo surgery by a gynecologic oncologist will continue to have the more extensive procedure. Interestingly, a recent report by Bryant et al. noted a significantly higher rate of complete surgical staging in white patients compared to African American patients (49 % vs. 38 %; $p = 0.001$) (Bryant et al. 2009).

6.3.3 Staging of Unstaged Malignant Ovarian Germ Cell Tumors

Not infrequently, patients are referred after having undergone primary surgery without comprehensive surgical staging. Unfortunately, scant information exists to guide gynecologic oncologists in managing this situation. In making a decision about reoperation, a relevant question is whether an additional procedure will change management or influence prognosis. Here again, the pediatric intergroup study would suggest that a second operative procedure is not indicated. In a recent report from the Multicenter Italian Trials in Ovarian Cancer (MITO) group, Mangili et al.

described 26 patients with pure ovarian dysgerminoma, of whom only five patients had complete surgical staging (Mangili et al. 2011). Three patients (11.5 %) relapsed; none of them had received adjuvant chemotherapy, and all had not received a complete surgical staging. However, all relapsed patients were salvaged. The authors concluded that, although conservative surgery with complete surgical staging is the gold standard, options for patients with incomplete staging could be surgical restaging or surveillance.

Our general approach to this clinical scenario is to perform imaging studies. If the ovarian mass has been resected, and no other abnormalities exist, then we would advise against surgery. If staging surgery is recommended, then a minimally invasive surgical approach is preferable, unless otherwise contraindicated. If imaging shows bulky residual disease, consideration will be given to a second procedure to achieve minimal residual disease, but in the absence of bulky disease, surgery is generally avoided. These recommendations differ somewhat from those for unstaged epithelial ovarian cancer, as germ cell tumors tend to be much more chemosensitive than their epithelial counterparts.

6.3.4 Primary Cytoreductive Surgery for Advanced Stage Disease

Whether we should follow the same principles that we apply in epithelial ovarian cancers to the management of these chemosensitive tumors remains unresolved. Little information to guide practice exists. Two reports from the Gynecologic Oncology Group (GOG) suggest a benefit from minimal residual disease at completion of primary cytoreductive surgery followed by either non-platinum- or platinum-based chemotherapy regimens (Slayton et al. 1985; Williams et al. 1989), but patients in these studies were not prescribed BEP as is commonly used for germ cell tumors today. While maximum cytoreduction to achieve minimal residual disease is the primary goal, one should also consider patient safety and

reducing morbidity in view of the excellent chemosensitivity of this disease spectrum.

6.4 Postoperative Therapy

6.4.1 Chemotherapy for Nondysgerminomatous Tumors

Prior to the advent of combination chemotherapy in the 1970s, the prognosis for patients with malignant ovarian germ cell tumors was poor. With the exception of patients with dysgerminoma, the survival rate for patients with apparent stage I disease was only 5–20 %, and the survival for those with advanced stage disease was negligible (Kurman and Norris 1976a, b, c, 1977; Jimerson and Woodruff 1977a, b; Gershenson et al. 1983, 1984, 1986a, b). In the 1970s, the combination of vincristine, dactinomycin, and cyclophosphamide (VAC) became the standard (Slayton et al. 1985; Gershenson et al. 1985). Although this regimen was moderately effective for patients with apparent early stage disease, it was relatively ineffective for those with advanced stage tumors. Since about the mid-1980s, the combination of bleomycin, etoposide, and cisplatin (BEP) has been the standard (Gershenson et al. 1990; Williams et al. 1994). Cure rates for patients with stage I disease approach 100 % compared with 75 % + for those with advanced stage tumors.

6.4.2 Chemotherapy for Dysgerminoma

Historically, the major distinction between pure dysgerminoma and the other subtypes of malignant ovarian germ cell tumors has been its exquisite radiosensitivity. Prior to the modern chemotherapy era, dysgerminoma was associated with a very high cure rate when treated with radiotherapy. However, in the 1980s, clinical experience and subsequent clinical trials clearly indicated that dysgerminoma is equally sensitive to cytotoxic chemotherapy (Gershenson et al.

Table 6.1 Surveillance following primary surgery

Study	No. pts.	Stages	Tumor types	Relapse	Outcome
Bonazzi et al. (1994)	22	I, II	IT—G1 & 2	2 (9 %)	IT G0: both salvaged with surgery alone (? growing teratoma syndrome)
Mitchell et al. (1999)	9	I	IT—G2, unknown grade Mixed GCT—all with YST	1 (11 %)	IT unknown grade; Refused further treatment
Cushing et al. (1999)	44	I (clinical)	IT—G1,2, 3 Mixed GCT—IT + YST	1 (2.3 %)	Salvaged with chemotherapy
Patterson et al. (2008)	36	I	IT—G0, 1, 2, & 3 Dysgerminoma Yolk sac tumor Embryonal carcinoma Mixed GCT	11 (30 %)	All salvaged with chemotherapy except for 1 pt. with high -grade IT who died of progressive disease and 1 pt. with G2 IT who died of PE while receiving chemotherapy
Mangili et al. (2010)	19	I	IT—G1, 2, 3	4 (21 %)	2 with IT G0 salvaged with surgery; 2 with IT salvaged with surgery + chemotherapy

Reports of surveillance strategy in patients with malignant ovarian germ cell tumors
Abbreviations: IT immature teratoma, *GCT* germ cell tumor, *YST* yolk sac tumor, *PE* pulmonary embolism

1986; Williams et al. 1991; Brewer et al. 1999; Williams et al. 2004). Of course, the major advantage of chemotherapy is the much lower incidence of associated premature ovarian failure.

6.4.3   Surveillance

Historically, the only patients considered to be candidates for treatment with surgery alone were those with stage IA pure dysgerminoma and stage IA, grade 1 immature teratoma. However, over the past few years, a number of groups have reported provocative results with the strategy of surveillance following primary surgery (Table 6.1). In considering this strategy, the critical issue is ultimate outcome. Of course, if surveillance is successful, one can avoid the myriad acute effects of chemotherapy as well as the risk of significant late effects, such as secondary leukemia or premature ovarian failure. Even if there is a moderate relapse risk necessitating chemotherapy, if one can avoid chemotherapy in a significant proportion of patients and concomitantly salvage those who relapse, such management will be deemed successful. On the other hand, if very many of

those who relapse ultimately succumb to their disease, it will most certainly not become part of standard practice. Where the threshold lies on the latter remains unresolved.

Bonazzi et al. conducted a prospective study of 22 patients with stage I or II, grade 1 or 2, immature teratoma of the ovary who were treated with surgery alone (Bonazzi et al. 1994). Two patients relapsed—one with stage IA, grade 2 and the other with stage IC, grade 2. Both were salvaged with surgery that revealed only grade 0 immature teratoma. Dark et al. reported on 24 patients who underwent surveillance after surgery for stage IA malignant ovarian germ cell tumors (Dark et al. 1997). Nine had dysgerminoma (which most already would observe); nine had immature teratoma; and six had yolk sac tumors, with or without immature teratoma. In a follow-up report from the Charing Cross/Mt. Vernon group in UK, which apparently included those reported earlier, 37 patients with stage I malignant ovarian germ cell tumors underwent surveillance (Patterson et al. 2008). However, one patient had a squamous carcinoma arising in a mature cystic teratoma, which, in my opinion, is a completely different entity. The other 36 patients included three with grade 0 immature teratoma (which would never require chemotherapy unless thought to represent a growing teratoma syndrome), nine dysgerminomas, one embryonal carcinoma, 15 immature teratomas, six yolk sac tumors, and two with mixed germ cell tumors. Of these 36 patients, 11 (31 %) relapsed—one with mature teratoma (which may represent growing teratoma syndrome), two with dysgerminoma (although one was a second primary in the contralateral ovary), four with immature teratoma, two with mixed germ cell tumors, and one each with yolk sac tumor and embryonal carcinoma. Two of the 11 patients died. One had grade 2 immature teratoma and became pregnant. She was diagnosed 13 months after diagnosis and died of a pulmonary embolus during chemotherapy. The second patient had grade 3 immature teratoma. She relapsed 3 months after diagnosis and subsequently died after secondary surgery and multiple lines of chemotherapy.

Mitchell et al. (1999) reported nine patients with stage I malignant ovarian germ cell tumors—six with immature teratoma and three with mixed germ cell tumors, all of whom had yolk sac tumor elements. At the time of the report, only one patient had relapsed. She had immature teratoma of unknown grade and relapsed 8 months after a unilateral salpingo-oophorectomy, but her parents refused further therapy. Cushing et al. (1999) described 44 patients with completely resected immature teratoma (31 pure and 13 with yolk sac tumor) treated with surgery alone. With a median follow-up of 4.2 years, one patient with a mixed germ cell tumor relapsed at 18 weeks, as indicated by a rising alpha fetoprotein, and was salvaged with 4 cycles of BEP. At the time of the report, she was disease-free at 57 months after completing chemotherapy.

Most recently, Mangili et al. (2010) reported 28 patients with stage I pure immature teratoma of whom 19 were treated with surgery alone. Of these 19 patients, nine had grade 1 disease, eight had grade 2, and two had grade 3 disease. Four patients (21 %) relapsed—two with mature teratoma and two with immature teratoma. The two with mature teratoma recurrence were treated with secondary cytoreductive surgery, and the two with immature teratoma were treated with secondary surgery followed by BEP chemotherapy. All were salvaged.

The Children's Oncology Group (COG) is currently studying the treatment strategy of surgery plus surveillance in children with stage I malignant ovarian germ cell tumors. If the findings appear to be promising, further study will be required in adult patients.

6.5 Treatment of Recurrent Malignant Ovarian Germ Cell Tumor

6.5.1 Secondary Cytoreductive Surgery

As in all types of ovarian malignancies, the role of secondary cytoreductive surgery remains unclear. There are retrospective reports that suggest a benefit of secondary debulking in selected patients with recurrent malignant ovarian germ cell tumors (Messing et al. 1992; Munkarah et al. 1994; Li et al. 2007). However, there are no prospective trials nor is there likely to be given the rarity of this tumor type and clinical scenario. Munkarah et al. (1994) reported that the survival of patients with recurrent immature teratoma who underwent secondary cytoreductive surgery was superior to those with other subtypes. Similarly, Li et al. (2007) reported that in chemorefractory patients with recurrent disease, patients who underwent secondary cytoreductive surgery for dysgerminoma or immature teratoma did better than patients with other histologic subtypes, as did patients who underwent optimal cytoreduction. In the context of inadequate data, ideal candidates for secondary debulking are those with an isolated focus or limited foci of recurrent tumor. Those with carcinomatosis, multiple hepatic metastases, or extensive extraperitoneal disease are unlikely to benefit from this therapeutic approach.

6.5.2 Systemic Treatment of Recurrence

Fortunately, recurrence of a malignant ovarian germ cell tumor is very uncommon. Of those patients who do recur, most do so within 24 months of diagnosis (Messing et al. 1992). Because of the rarity of relapse in this disease, no standard therapy exists. Consequently, treatment strategies have been extrapolated from clinical experience in testicular cancer. Currently, the most common salvage regimens consist of either the combination of paclitaxel, ifosfamide, and cisplatin (TIP) or the combination of vinblastine, ifosfamide, and cisplatin (VIP) (Loehrer et al. 1988, 1998; McCaffrey et al. 1997; Kondagunta et al. 2005). The VIP or TIP regimen is generally recommended as an induction strategy for 1–2 cycles followed by high-dose chemotherapy with stem cell rescue, again predominantly based on experience in testicular cancer (Broun et al. 1997; Bhatia et al. 2000; Motzer et al. 2000; Einhorn et al. 2007). Of note, a better prognosis was seen among patients with chemorefractory disease who received standard BEP/PVB salvage chemotherapy non-BEP/PVB primary chemotherapy, potentially highlighting the superior efficacy of this treatment regimen (Li et al. 2007).

6.6 Growing Teratoma Syndrome

The growing teratoma syndrome is extremely rare (Kattan et al. 1993; Amsalem et al. 2004; Zagame et al. 2006). Diagnostic criteria include the following: (1) clinical or radiological enlargement of tumor masses during or after chemotherapy administered for a malignant ovarian germ cell tumor that originally has immature teratoma elements; (2) normalization of previously elevated serum tumor markers—AFP, hCG, or both; and (3) only mature (grade 0) teratoma without any malignant germ cell component in the resected tissue. Because of the tenfold difference in incidence, growing teratoma syndrome is much more common in males. Treatment consists of complete surgical resection, whenever feasible. Zagame et al. reported 12 patients with growing teratoma syndrome (Zagame et al. 2006). The median interval between the diagnosis of an immature teratoma of the ovary and the growing teratoma syndrome was 9 months, with a range of 4–55 months. In nine of the 12 cases, the diagnosis was made on radiological examination. Complete surgical resection was performed in eight patients, and all but one patient, who was lost to follow-up, were alive at the time of the report.

References

Amsalem H, Nadjari M, Prus D et al (2004) Growing teratoma syndrome vs. chemotherapeutic retroconversion. Case report and review of literature. Gynecol Oncol 92:357–360

Beiner ME, Gotlieb WH, Korach Y et al (2004) Cystectomy for immature teratoma of the ovary. Gynecol Oncol 93:381 384

Bhatia S, Abonour R, Porcu P et al (2000) High-dose chemotherapy as initial salvage chemotherapy in patients with relapsed testicular cancer. J Clin Oncol 18: 3346–3351

Billmire D, Vinocur C, Rescorla F et al (2004) Outcome and staging evaluation in malignant germ cell tumors of the ovary in children and adolescents: an intergroup study. J Pediatr Surg 39:424–429; discussion 429

Bonazzi C, Peccatori F, Colombo N et al (1994) Pure ovarian immature teratoma, a unique and curable disease: 10 years' experience of 32 prospectively treated patients. Obstet Gynecol 84:598–604

Brewer M, Gershenson DM, Herzog CE et al (1999) Outcome and reproductive function after chemotherapy for ovarian dysgerminoma. J Clin Oncol 17:2670–2675

Broun ER, Nichols CR, Gize G et al (1997) Tandem high dose chemotherapy with autologous bone marrow transplantation for initial relapse of testicular germ cell cancer. Cancer 79:605–610

Bryant CS, Kumar S, Shah JP (2009) Racial disparities in survival among patients with germ cell tumors of the ovary—United States. Gynecol Oncol 114:437–441

Cushing B, Giller R, Ablin A et al (1999) Surgical resection alone is effective treatment for ovarian immature teratoma in children and adolescents: a report of the Pediatric Oncology Group and the Children's Cancer Group. Am J Obstet Gynecol 181:353–358

Dark GG, Bower M, Newlands ES et al (1997) Surveillance policy for stage I ovarian germ cell tumors. J Clin Oncol 15:620–624

Einhorn LH, Williams SD, Chamness A et al (2007) High-dose chemotherapy and stem-cell rescue for metastatic germ-cell tumors. N Engl J Med 357: 340–348

Gershenson DM, Del Junco G, Herson J et al (1983) Sinus tumor of the ovary: the M.D. Anderson experience. Obstet Gynecol 61:194–202

Gershenson DM, Del Junco G, Copeland LJ et al (1984) Mixed germ cell tumors of the ovary. Obstet Gynecol 64:200–206

Gershenson DM, Copeland LJ, Kavanagh JJ et al (1985) Treatment of malignant nondysgerminomatous germ cell tumors of the ovary with vincristine, dactinomycin, and cyclophosphamide. Cancer 56:2756–2761

Gershenson DM, del Junco G, Silva EG et al (1986a) Immature teratoma of the ovary. Obstet Gynecol 68: 624–629

Gershenson DM, Wharton JT, Kline RC et al (1986b) Chemotherapeutic complete remission in patients with metastatic ovarian dysgerminoma. Potential for cure and preservation of reproductive capacity. Cancer 58: 2594–2599

Gershenson DM, Morris M, Cangir A et al (1990) Treatment of malignant germ cell tumors of the ovary with bleomycin, etoposide, and cisplatin. J Clin Oncol 8:715–720

Gershenson D, Miller A, Champion V et al (2007) Reproductive and sexual function after platinum-based chemotherapy in long-term ovarian germ cell tumor survivors: a Gynecologic Oncology Group study. J Clin Oncol 25(19):2792–2797

The International Prognostic Factors Study Group (2010) Prognostic factors in patients with metastatic germ cell tumors who experienced treatment failure with cisplatin-based first-line chemotherapy. J Clin Oncol 28:4906–4911

Jimerson GK, Woodruff JD (1977a) Ovarian extraembryonal teratoma: I. Endodermal sinus tumor. Am J Obstet Gynecol 127:73–79

Jimerson GK, Woodruff JD (1977b) Ovarian extraembryonal teratoma: II. Endodermal sinus tumor mixed with other germ cell tumors. Am J Obstet Gynecol 127: 302–305

Kattan J, Droz JP, Culine S et al (1993) The growing teratoma syndrome: a woman with nonseminomatous germ cell tumor of the ovary. Gynecol Oncol 49: 395–399

Kondagunta GV, Bacik J, Donadio A et al (2005) Combination of paclitaxel, ifosfamide, and cisplatin is an effective second-line therapy for patients with relapsed testicular germ cell tumors. J Clin Oncol 23: 6549–6555

Kumar S, Shah JP, Bryant CS et al (2008) The prevalence and prognostic impact of lymph node metastasis in malignant germ cell tumors of the ovary. Gynecol Oncol 110:125–132

Kurman RJ, Norris HJ (1976a) Malignant mixed germ cell tumors of the ovary: a clinical and pathologic analysis of 30 cases. Obstet Gynecol 48:579–589

Kurman RJ, Norris HJ (1976b) Embryonal carcinoma of the ovary: a clinicopathologic entity distinct from endodermal sinus tumor resembling embryonal carcinoma of the adult testis. Cancer 38:2420–2433

Kurman RJ, Norris HJ (1976c) Endodermal sinus tumor of the ovary: a clinical and pathologic analysis of 71 cases. Cancer 38:2404–2419

Kurman RJ, Norris HJ (1977) Malignant germ cell tumors of the ovary. Hum Pathol 8:551–564

Lai CH, Chang TC, Hsueh S et al (2005) Outcome and prognostic factors in ovarian germ cell malignancies. Gynecol Oncol 96:784–791

Li J, Yang W, Wu X (2007) Prognostic factors and role of salvage surgery in chemorefractory ovarian germ cell malignancies: a study in Chinese patients. Gynecol Oncol 105:769–775

Loehrer PJ Sr, Lauer R, Roth BJ et al (1988) Salvage therapy in recurrent germ cell cancer: ifosfamide and cisplatin plus either vinblastine or etoposide. Ann Intern Med 109:540–546

Loehrer PJ Sr, Gonin R, Nichols CR et al (1998) Vinblastine plus ifosfamide plus cisplatin as initial salvage therapy in recurrent germ cell tumor. J Clin Oncol 16:2500–2504

Mangili G, Scarfone G, Gadducci A et al (2010) Is adjuvant chemotherapy indicated in stage I pure immature ovarian teratoma (IT)? A multicentre Italian trial in ovarian cancer (MITO-9). Gynecol Oncol 119: 48–52

Mangili G, Sigismondi C, Lorusso D et al (2011) Is surgical restaging indicated in apparent stage IA pure ovarian dysgerminoma? The MITO group retrospective experience. Gynecol Oncol 121:280–284

McCaffrey JA, Mazumdar M, Bajorin DF et al (1997) Ifosfamide and cisplatin-containing chemotherapy as first-line salvage therapy in germ cell tumors; response and survival. J Clin Oncol 15:2559–2563

Messing MJ, Gershenson DM, Morris M et al (1992) Primary treatment failure in patients with malignant ovarian germ cell neoplasms. Int J Gynecol Cancer 2:295–300

Mitchell PL, Al-Nasiri N, A'Hern R et al (1999) Treatment of nondysgerminomatous ovarian germ cell tumors: an analysis of 69 cases. Cancer 85:2232–2244

Motzer RJ, Mazumdar M, Sheinfeld J et al (2000) Sequential dose-intensive paclitaxel, ifosfamide, carboplatin, and etoposide salvage therapy for germ cell tumor patients. J Clin Oncol 18:1173–1180

Munkarah A, Gershenson DM, Levenback C et al (1994) Salvage surgery for chemorefractory ovarian germ cell tumors. Gynecol Oncol 55:217–223

Murugaesu N, Schmid P, Dancey G et al (2006) Malignant ovarian germ cell tumors: identification of novel prognostic markers and long-term outcome after multimodality treatment. J Clin Oncol 24: 4862–4866

Norris HJ, Zirkin HJ, Benson WL (1976) Immature (malignant) teratoma of the ovary: a clinical and pathologic study of 58 cases. Cancer 37:2359–2372

Patterson DM, Murugaesu N, Holden L et al (2008) A review of the close surveillance policy for stage 1 female germ cell tumors of the ovary and other sites. Int J Gynecol Cancer 18(1):43–50

Roth LM, Talerman A (2006) Recent advances in the pathology and classification of ovarian germ cell tumors. Int J Gynecol Pathol 25:305–320

Slayton RE, Park RC, Silverberg SG et al (1985) Vincristine, dactinomycin, and cyclophosphamide in the treatment of malignant germ cell tumors of the ovary: a Gynecologic Oncology Group study (a final report). Cancer 56:243–248

Tavassoli FA, Deville P (2003) Pathology and genetics of tumours of the breast and female genital organs. International Agency for Research on Cancer, Lyon

Weinberg LE, Lurain JR, Singh DK et al (2011) Survival and reproductive outcomes in women treated for malignant ovarian germ cell tumors. Gynecol Oncol 121:285–289

Williams SD, Blessing JA, Moore DH et al (1989) Cisplatin, vinblastine, and bleomycin in advanced and recurrent ovarian germ-cell tumors: a trial of the Gynecologic Oncology Group. Ann Intern Med 111: 22–27

Williams SD, Blessing JA, Hatch KD et al (1991) Chemotherapy of advanced dysgerminoma: trials of the Gynecologic Oncology Group. J Clin Oncol 9: 1950–1955

Williams SD, Blessing JA, DiSaia PJ et al (1994) Second-look laparotomy in ovarian germ cell tumors: the Gynecologic Oncology Group experience. Gynecol Oncol 52:287–291

Williams SD, Kauderer J, Burnett AF et al (2004) Adjuvant therapy of completely resected dysgerminoma with carboplatin and etoposide: a trial of the Gynecologic Oncology Group. Gynecol Oncol 95: 496–499

Zagame L, Pautier P, Duvillard P et al (2006) Growing teratoma syndrome after ovarian germ cell tumors. Obstet Gynecol 108:509–514

Zanetta G, Bonazzi C, Cantu MG et al (2001) Survival and reproductive function after treatment of malignant germ cell ovarian tumors. J Clin Oncol 19: 1015–1020

Ovarian and Testicular Sex Cord-Stromal Tumors

7

Kris Ann P. Schultz, Lindsay Frazier,
and Dominik T. Schneider

Contents

K.A.P. Schultz, MD, MS (✉)
Department of Hematology and Oncology,
Children's Hospitals and Clinics of Minnesota,
Minneapolis, MN, USA
e-mail: krisann.schultz@childrensmn.org

L. Frazier, MD
Department of Pediatric Oncology,
Dana Farber Cancer Institute, Boston, MA, USA

D.T. Schneider, MD
Department of Pediatrics,
Klinikum Dortmund, Dortmund, Germany
e-mail: dominik.schneider@klinikumdo.de

7.1 Ovarian Sex Cord-Stromal Tumors

Ovarian tumors comprise 1 % of childhood cancer (Breen and Maxson 1977) and are the most frequent neoplasm of the female genital tract (Hassan et al. 1999). Ovarian sex cord-stromal tumors account for approximately 15 % of ovarian tumors in pediatrics and approximately 7 % of ovarian tumors in adults (Koonings et al. 1989). In young children, ovarian germ cell tumors are more common than ovarian stromal tumors. In adults, epithelial ovarian tumors predominate.

Ovarian sex cord-stromal tumors develop from the specific gonadal stroma that supports germ cell development and hormone secretion. The histologic differentiation is not restricted to the sex of the patient. Thus, ovarian sex cord-stromal tumors include juvenile granulosa cell tumors, adult granulosa cell tumors, Sertoli-Leydig cell tumors, Sertoli tumors, sclerosing stromal tumors, sex cord-stromal tumor with annular tubules, undifferentiated sex cord-stromal tumors, and gynandroblastoma. Gynandroblastomas contain both granulosa cell and Sertoli cell elements. The recently described microcystic stromal tumor of the ovary is likely to be grouped within the sex cord-stromal tumor category, although this tumor type has not yet been reported in a pediatric patient (youngest reported to date, 26 years) (Irving and Young 2009). Sclerosing stromal tumors may occur within pediatrics and are generally regarded as benign (Fefferman et al. 2003).

A.L. Frazier, J.F. Amatruda (eds.), *Pediatric Germ Cell Tumors*, Pediatric Oncology 1,
DOI 10.1007/978-3-642-38971-9_7, © Springer-Verlag Berlin Heidelberg 2014

In children and adolescents, juvenile granulosa cell tumors are the most common subtype, followed by Sertoli-Leydig cell tumors. The age at presentation of juvenile granulosa cell tumors is younger than that in Sertoli-Leydig cell tumors and primarily includes prepubertal patients. In contrast, the incidence of Sertoli-Leydig cell tumors increases during adolescence.

Extragonadal sex cord-stromal tumors may also occur, usually in adult women and within the pelvis, including the broad ligament. Typically these cases are either granulosa cell tumors or sex cord-stromal tumors with annular tubules histology.

7.1.1 Normal Ovarian Structures

In contrast to epithelial and germ cell tumors, sex cord-stromal tumors arise from the sex cord or ovarian stroma/mesenchyme of the developing gonad. Embryologically, ovarian sex cords arise from the primitive genital ridge and become the cortical cords. Normally, the sex cords develop into the ovarian follicles (Satoh 1991).

7.1.2 Clinical Presentation

As noted in other ovarian tumors, ovarian sex cord-stromal tumors may present with abdominal pain or distention (Schultz et al. 2005). In contrast to germ cell or epithelial ovarian tumors, however, ovarian sex cord-stromal tumors may also present with signs of hormone production such as precocious puberty, or in older girls primary or secondary amenorrhea or virilization (Schultz et al. 2005). Ovarian sex cord-stromal tumors show staining for sex steroid biosynthesis enzymes, thus androgenic clinical findings are likely related to synthesis of hormones by sex steroid biosynthesis enzymes present within tumor tissue (Costa et al. 1994).

Juvenile granulosa cell tumors typically present in very young girls with signs of precocious puberty or abdominal distention (Schneider et al. 2003a). In some patients, tumors may lead to ovarian torsion or spontaneous tumor rupture; these patients clinically present with acute abdominal pain. Juvenile granulosa cell tumors may present in older girls with primary or secondary amenorrhea. When found at an early stage (International Federation of Gynecology and Obstetrics [FIGO] stage Ia), these tumors may be treated with resection alone and generally carry a favorable prognosis (Schneider et al. 2003a).

When found at a later stage, juvenile granulosa cell tumors carry a poorer prognosis, especially when associated with high mitotic index (see Sect. 7.1.7). It is critical to distinguish juvenile granulosa cell tumors from the adult granulosa cell tumors usually seen in older women. Adult granulosa cell tumors carry a significant risk for late recurrence, which is not the case in juvenile granulosa cell tumors.

Sertoli-Leydig cell tumors may present with abdominal pain or mass or with signs of virilization including hirsutism and acne.

Sex cord-stromal tumor with annular tubules both within and outside the setting of Peutz-Jeghers syndrome (PJS) may present with estrogenic manifestations (Gibbon 2005).

7.1.3 Differential Diagnosis

7.1.3.1 Germ Cell Tumors

In the absence of signs of hormone production, the clinical presentation of ovarian germ cell tumors and ovarian stromal tumors may be identical. Measurement of serum alpha fetoprotein (AFP) and beta human chorionic gonadotropin (HCG) is advised for all girls with a possible ovarian mass. AFP may be elevated in some cases of Sertoli-Leydig cell tumor, particularly Sertoli-Leydig cell tumors with reticular pattern. Carcinoembryonic antigen (CEA) measurement is optional, but may be elevated in some cases of mucinous ovarian tumors. CEA elevation has also been described in adults with adenocarcinoma components arising within a mature teratoma (Sumi et al. 2002).

7.1.3.2 Female Adnexal Tumor of Wolffian Origin

This tumor usually arises within the broad ligament but may appear to be an ovarian mass. Histopathologically, this lesion may strongly resemble Sertoli-Leydig cell tumor (Sivathondan et al. 1979).

7.1.3.3 Epithelial Ovarian Tumors

During adolescence the risk of developing epithelial ovarian tumors gradually increases. This includes cystadenomas, both serous and mucinous; borderline tumors; and invasive ovarian adenocarcinoma. The association with BRCA gene mutations in this age group remains to be investigated. Most tumors present as stage I.

7.1.3.4 Small Cell Ovarian Carcinoma

Sex cord-stromal tumors may also closely resemble small cell ovarian carcinoma of the hypercalcemic type. Close attention to serum calcium levels at the time of clinical presentation is highly recommended (Young 2005).

Inhibin antibody may also be helpful in differentiating ovarian sex cord-stromal tumors from small cell ovarian carcinoma. Positive staining with inhibin antibody may distinguish small cell carcinoma from juvenile granulosa cell tumor and Sertoli-Leydig cell tumor, although it cannot distinguish between the latter two diagnoses (Rishi et al. 1997; Zheng et al. 2003). Small cell carcinoma carries a dismal prognosis if aggressive treatment is not pursued. This entity is discussed in detail in Rare Tumors in Children and Adolescents (Schneider et al. 2012).

7.1.4 Genetics

7.1.4.1 Hereditary Predisposition

Juvenile granulosa cell tumors are generally seen in otherwise healthy children with no evidence of hereditary predisposition but may also occur in children with multiple enchondromatosis (Ollier's disease), Maffucci syndrome, and PJS (Vaz and Turner 1986; Velasco-Oses et al. 1988; Tanaka et al. 1992).

Ovarian sex cord-stromal tumors with annular tubules are often seen in conjunction with PJS. In fact, one-third of patients with these tumors also have PJS; these patients tend to present at a younger age and have a higher incidence of bilateral disease. Peutz-Jeghers syndrome

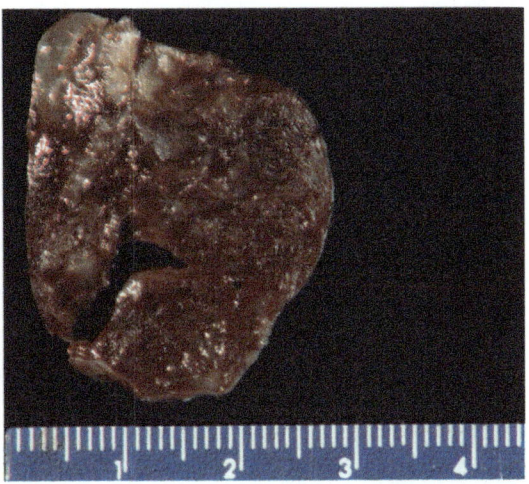

Fig. 7.1 Gross specimen of a Sertoli cell tumor of the ovary in a 6-year-old female with a previous history of multiple pleuropulmonary blastomas and a cystic nephroma; the child presented with persistent abdominal pain. The hemorrhagic appearance of this 8 cm mass reflects torsion with hemorrhagic infarction. Photograph provided by D. Ashley Hill, MD and Louis P. Delmer, MD. Reprinted with permission from Elsevier (Schultz et al. 2011)

is associated with mutations in STK11/LKB1 (chromosome 19p13.3), although the role of this gene in the development of ovarian tumors is unclear (Beggs Beggs et al. 2010).

A 2004 study of sporadic sex cord-stromal tumors showed loss of heterozygosity at 19p13.3 without mutations or promotor methylation of STK11 (Kato et al. 2004).

Ovarian Sertoli-Leydig cell tumors, juvenile granulosa cell tumors, and gynandroblastomas have been noted in families of children with pleuropulmonary blastoma, a rare lung tumor of childhood (Figs. 7.1 and 7.2) (Schultz et al. 2011). DICER1 (located on chromosome 5) mutations are seen in the majority of children with pleuropulmonary blastoma (Hill et al. 2009) and have also been documented in children with Sertoli-Leydig cell tumor, juvenile granulosa cell tumor, and gynandroblastoma (Schultz et al. 2011). DICER1 mutations may also mediate the association of Sertoli-Leydig cell tumors with multinodular goiter (Heravi-Moussavi et al. 2012; Rio Frio et al. 2011; Schultz et al. 2011; Slade et al. 2011).

Androgen insensitivity syndrome has been associated with Sertoli-Leydig cell tumors

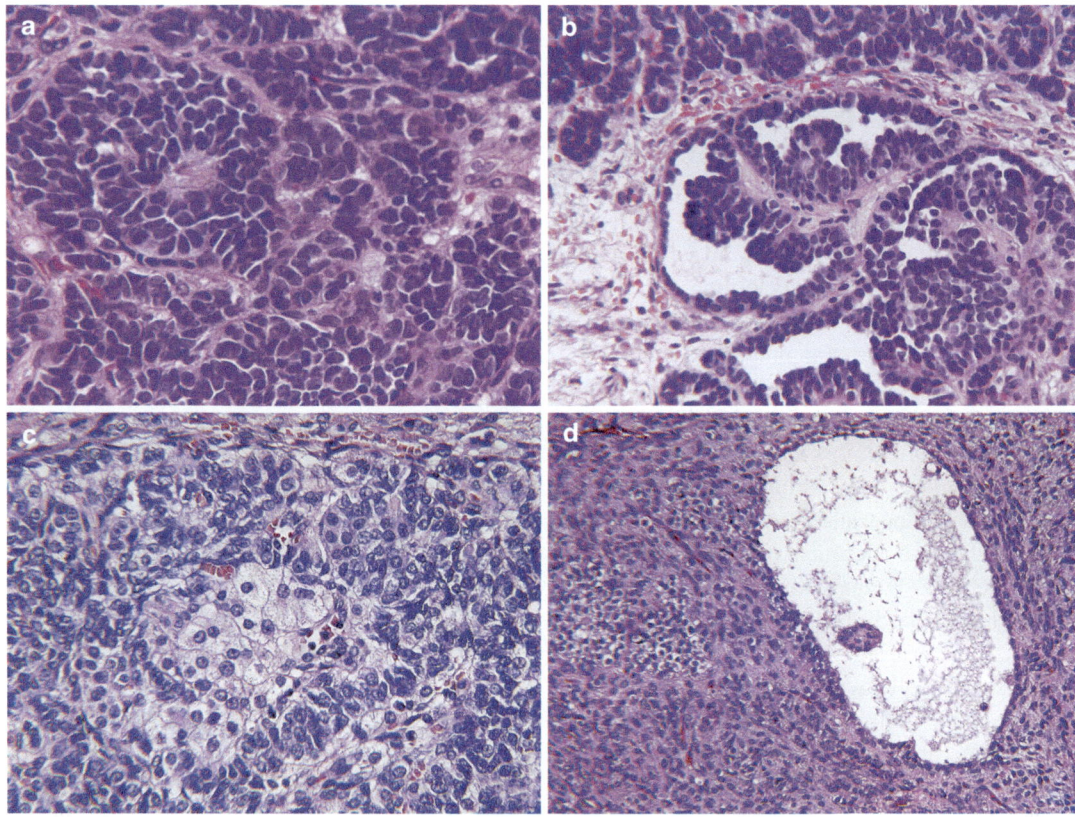

Fig. 7.2 (**a**) Sertoli cell tumor of the ovary showing the pattern of solid cords of uniform basophilic cells. (**b**) Other patterns in the same tumor demonstrate small tubules adjacent to retiform or papillary profiles. (**c**) Sertoli-Leydig cell tumor, largely necrotic and hemorrhagic. Viable areas show the presence of small basophilic cells in cord-like structures with intermixed clusters of larger, pale staining cells representing Leydig cells. (**d**) Juvenile granulosa cell tumor with a cystic structure resembling a graffian follicle surrounded by a densely cellular stroma. Photomicrographs provided by (**d**). Ashley Hill, MD and Louis P. Dehner, MD. Reprinted with permission from Elsevier (Schultz et al. 2011).

(Ozülker et al. 2010; Stacher et al. 2010; Rutgers 2011). Adult granulosa cell tumors have been associated with a recurring somatic mutation in FOXL2 (Shah et al. 2009; Jamieson et al. 2010; Kim and Sung et al. 2010b). Additionally, fibromas may be seen in the basal cell nevus syndrome (Ball et al. 2011). Although epithelial ovarian tumors may be associated with BRCA1 or BRCA2 mutations, no such association has been described in ovarian stromal tumors.

7.1.4.2 Tumor Cytogenetics
Although the majority of these tumors show a balanced karyotype, 25 % show an unbalanced karyotype—usually a gain of chromosome 12, which is of unclear clinical significance (Schneider et al. 2003b).

7.1.5 Recommended Evaluations

Attention during the history and physical examination should be directed towards the presence of any gastrointestinal or systemic symptoms and the presence of thyroid nodules, as well as the presence of any endocrine manifestations such as signs of isosexual precocity, virilization (hirsutism, acne, clitoromegaly, voice changes), or menstrual irregularities.

Table 7.1 FIGO staging criteria for ovary

Carcinoma of the ovary; FIGO nomenclature (Rio de Janeiro 1988)		
Stage I	Growth limited to the ovaries	
	Ia	Growth limited to one ovary; no ascites present containing malignant cells. No tumor on the external surface; capsule intact
	Ib	Growth limited to both ovaries; no ascites present containing malignant cells. No tumor on the external surfaces; capsules intact
	Ic[a]	Tumor either Stage 1a or 1b, but with tumor on surface of one or both ovaries, or with capsule ruptured, or with ascites present containing malignant cells, or with positive peritoneal washings
Stage II	Growth involving one or both ovaries with pelvic extension	
	IIa	Extension and/or metastases to the uterus and/or tubes
	IIb	Extension to other pelvic tissues
	IIc[a]	Tumor either Stage IIa or IIb, but with tumor on surface of one or both ovaries, or with capsule(s) ruptured, or with ascites present containing malignant cells, or with positive peritoneal washings
Stage III	Tumor involving one or both ovaries with histologically confirmed peritoneal implants outside the pelvis and/or positive regional lymph nodes. Superficial liver metastases equals Stage III. Tumor is limited to the true pelvis, but with histologically proven malignant extension to small bowel or omentum	
	IIIa	Tumor grossly limited to the true pelvis, with negative nodes, but with histologically confirmed microscopic seeding of abdominal peritoneal surfaces, or histologic proven extension to small bowel or mesentery
	IIIb	Tumor of one or both ovaries with histologically confirmed implants, peritoneal metastasis of abdominal peritoneal surfaces, none exceeding 2 cm in diameter; notes are negative
	IIIc	Peritoneal metastasis beyond the pelvis >2 cm in diameter and/or positive regional lymph nodes
Stage IV	Growth involving one or both ovaries with distant metastases. If pleural effusion is present, there must be positive cytology to allot a case to Stage IV. Parenchymal liver metastasis equals Stage IV	

IJGO Vol. 95, Suppl. 1; FIGO Annual Report, Vol. 26
Reprinted with permission from Elsevier (FIGO Committee on Gynecologic Oncology (2009)
[a]In order to evaluate the impact on prognosis of the different criteria for allotting cases to Stage Ic or IIc, it would be of value to know if rupture of the capsule was spontaneous, or caused by the surgeon; and if the source of malignant cells detected was peritoneal washings, or ascites

Although several genetic syndromes are associated with the development of ovarian sex cord-stromal tumors, most children diagnosed with these uncommon tumors have no known predisposing factors. All patients should be carefully interviewed and examined with attention to any personal or family history of lentigines, growths, cancerous or dysplastic conditions, and thyroid disease, especially hyperplastic thyroid disease. Attention to the family history and presence of thyroid nodules may also reveal evidence for underlying *DICER1* mutations (Rio Frio et al. 2011).

Laboratory evaluations should include the measurement of AFP, beta-HCG, CEA, CA-125, LDH, inhibin A and B, and serum calcium.

7.1.6 Staging

Surgical staging is necessary for adequate treatment of ovarian sex cord-stromal tumors (see Table 7.1).

In pediatrics and young adult patients, preservation of fertility and ovarian function is favored when feasible. Salpingoophorectomy with staging procedures including palpation of the contralateral ovary is the initial surgery considered most reasonable for most patients. When metastatic disease is encountered, general principles of cytoreductive surgery should be followed (Gershenson 1994). Nevertheless, hysterectomy is generally not recommended even in metastatic

Fig. 7.3 Event-free survival of 54 patients with ovarian sex cord-stromal tumors correlated with mitotic activity of the tumor. Reprinted with permission from the American Society of Clinical Oncology (Schneider et al. 2003a)

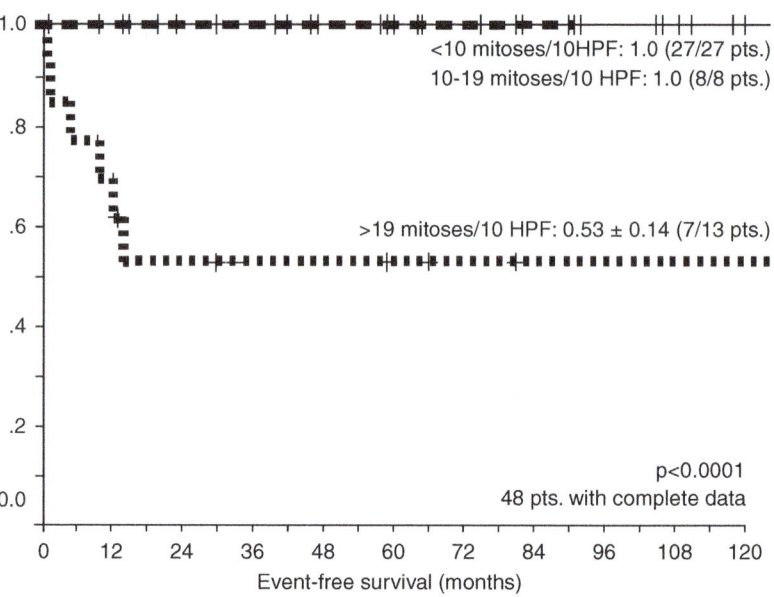

tumors when fertility preservation is a significant concern.

Given the rarity of lymph node involvement at diagnosis, routine staging lymphadenectomy may be reasonably omitted. However, lymph nodes which are enlarged by imaging or intraoperative examination should be removed (Brown et al. 2009).

Cytologic examination of peritoneal fluid or peritoneal washings is a critical part of the staging evaluation and must be achieved in every patient with an ovarian tumor. Careful attention to issues of preoperative versus intraoperative rupture is necessary given the relevance of rupture to prognosis and treatment (Schneider et al. 2003a).

Venous sampling for the diagnosis of a small Sertoli-Leydig cell tumor has recently been attempted with success. White et al. described its use in a 14-year-old patient presenting with signs of virilization, increased testosterone, and inconclusive ultrasound findings. Laparoscopic sampling of the ovarian vein conclusively identified the right ovary as the source of the increased testosterone (White et al. 2003).

Case reports and anecdotal experience suggest Sertoli-Leydig cell tumors may be FDG avid, but the frequency of PET avidity is not clear (Ozulker et al. 2010).

7.1.7 Prognostic Factors

Schneider et al. collected 72 cases of ovarian sex cord-stromal tumors over 20 years. Outcome correlated with FIGO stage and mitotic activity. For tumors with <20 versus ≥20 mitoses per high power field, event-free survival was 1.0 versus 0.48, respectively ($p = 0.0001$) (Schneider et al. 2003a, also see Fig. 7.3). However, if tumors were completely restricted to the ovary with no signs of microscopic spread, no recurrences were observed even in tumors with a high mitotic rate.

Within the category of Sertoli-Leydig cell tumors, poor differentiation and presence of heterologous elements are poor prognostic factors and generally indicate the need for adjuvant therapy for tumors beyond stage Ia.

Current data of the German MAKEI group indicate that in the case of juvenile granulosa cell tumors, intraoperative violation or rupture of the tumor does not affect prognosis. In contrast, spontaneous preoperative rupture indicates an increased risk of relapse—comparable to tumors with metastatic spread. However, it should be noted that in Sertoli-Leydig cell tumors, the risk of relapse may rise after intraoperative violation of the tumor, particularly if the tumor is poorly differentiated or shows a retiform pattern (Dominik Schneider, 2011, unpublished data; personal communication).

7.1.8 Treatment

Given the rarity of these diagnoses, prospective randomized trial results are not available to guide treatment decisions. Standard approaches are guided by small case series.

Despite limited prospective data, retrospective analysis suggests that early aggressive therapy in tumors other than stage I is critical. Relapse is associated with a low survival rate.

For young girls with completely resected FIGO stage Ia juvenile granulosa cell tumors (no preoperative rupture), observation alone is a reasonable strategy, as data suggest only minimal, if any, risk of relapse in this group (Merras-Salmio et al. 2002).

For patients with advanced stage ovarian juvenile granulosa cell tumor (>stage Ic), cure is possible but requires an aggressive approach. Previous reports suggest a low survival rate, although Schneider et al. describe 6/7 patients alive (4/7 in continuous first remission) after an aggressive regimen of 4–6 cycles of cisplatin-based therapy. One patient with peritoneal metastases also received 40 Gy abdominal irradiation (Schneider et al. 2002).

Treatment is most controversial for patients with tumors categorized as FIGO stage Ic due to intraoperative versus preoperative tumor rupture. In general, patients with only minor tumor violation during surgery can be followed expectantly and without adjuvant chemotherapy. In contrast, adjuvant cisplatin-based chemotherapy is recommended if tumors have ruptured spontaneously prior to surgery, or if peritoneal washings were positive. For patients with these tumors, four cycles of chemotherapy are most commonly recommended.

If patients show metastatic spread within the pelvis (FIGO II) or peritoneal cavity (FIGO III), the number of chemotherapy cycles may be increased to up to six cycles. After 2 cycles, patients should be restaged, and any residual tumors should be resected after four cycles of chemotherapy.

For patients with metastatic disease and histologic features indicating unfavorable prognosis (such as high mitotic rate or Sertoli-Leydig histology with poor differentiation), therapy intensification should be discussed. This may include high-dose chemotherapy with autologous stem cell transplantation, abdominal radiotherapy, or thermochemotherapy, based on the individual's prior treatment. However, it should be remembered that such strategies have not been validated by prospective trials and represent an individual approach for the patient whose prognosis is considered poor if given standard treatment.

Gershenson et al. reported the activity of cisplatin, etoposide, and bleomycin in a group of 9 adult patients with metastatic ovarian sex cord-stromal tumors or stage I poorly differentiated Sertoli-Leydig cell tumors. Although the response rate was 83 %, only one of the patients with metastatic disease had a durable remission, leading the authors to conclude that additional agents were needed to provide a durable response (Gershenson et al. 1996). Favorable activity of platinum and taxane-based chemotherapy on sex cord-stromal tumors has also reported (Brown et al. 2005).

The role of radiation therapy in the treatment of sex cord-stromal tumors is not clear. Ramirez et al. studied the use of letrozole in estrogen-positive, high-grade ovarian and peritoneal malignancies which were resistant to platinum and taxane-based therapy. Twenty-six percent of patients showed stable disease or had a partial response (Ramirez et al. 2008). Since some ovarian sex cord-stromal tumors express estrogen receptors, letrozole may provide a palliative option for some late-stage patients.

7.1.9 Follow-Up

Depending on the stage, histology, and patient age, ultrasound, computed tomography, and magnetic resonance imaging (MRI) have all been used to follow patients with ovarian stromal tumors during and after therapy. Use of MRI has a clear advantage because it does not expose patients to radiation, but is limited by the potential need for sedation in young girls. Follow-up duration should be determined by considering the usual time to recurrence. Most patients who relapse with Sertoli-Leydig cell tumors will do so within 2–3 years of diagnosis, although later relapses have been described. Juvenile granulosa cell tumors are less likely to recur but will generally do so within a similar time period. In contrast, adult granulosa cell tumors are associated with late relapse, 10 or more years after diagnosis.

In Sertoli-Leydig cell tumors, there is a risk of metachronous contralateral tumors that may arise as late as 5 years after diagnosis. Therefore, these patients should undergo extended follow-up that should include screening for thyroid disease, since a small proportion of patients may develop goiter or thyroid carcinoma (Dominik Schneider, 2011, personal communication). The relationship of these to germline mutations in *DICER1* is a topic of ongoing research.

Tumor markers often play an important role in the follow-up of these patients. If elevated at diagnosis, tumor markers should be followed closely during the primary risk period for recurrence, with increasing intervals thereafter. The necessary duration of follow-up is not entirely clear, but is anticipated to be less than 10 years for those with Sertoli-Leydig or juvenile granulosa cell tumors and greater than 10 years for those with adult granulosa cell tumors.

7.2 Stromal Testis Tumors

7.2.1 Background

The vast majority of testicular tumors are of germ cell origin, although testicular sex cord-stromal tumors occur in prepubertal and postpubertal males. Testicular sex cord-stromal tumors originate from either Sertoli or Leydig cells but may also present as juvenile granulosa cell tumors, a pattern resembling ovarian histology.

7.2.2 Normal Structures of the Testis

Sertoli cells are stimulated by follicle-stimulating hormone and produce inhibin. Aromatase from Sertoli cells helps to convert testosterone to 17-beta-estradiol to direct spermatogenesis. Dicer, the enzyme encoded by *DICER1*, is required for normal Sertoli cell function and survival in a murine model (Kim et al. 2010a). Leydig cells are located between the seminiferous tubules and produce testosterone when stimulated by luteinizing hormone. Leydig cell neoplasms are the most common stromal testis tumor found beyond infancy.

7.2.3 Clinical Presentation

Prepubertal testis tumors are rare, accounting for only 1 % of pediatric solid tumors. Postpubertal testicular tumors, usually of germ cell subtype, are considerably more common and are considered elsewhere in this text.

The primary data on boys younger than 12 with testicular tumors comes from the Prepubertal Testis Tumor Registry, previously open through the American Academy of Pediatrics Urology Section. Data obtained from this Registry suggests that most testis tumors in young children behave in a benign fashion, with the exception of yolk sac tumors and undifferentiated stromal cell tumors (Ross 2002). Sertoli cell tumors in older boys may behave in a more malignant fashion.

Nearly all testis tumors in childhood present with a painless testicular mass that is usually easily palpable and seen on ultrasound as a heterogeneous solid mass. Boys may present with signs of hormone production such as gynecomastia or precocious puberty. In contrast to other testicular tumors, Leydig cell tumors may present with signs of hormone production without palpable mass. In these cases, ultrasound shows a heterogeneous testicular mass.

Unlike young girls with juvenile granulosa cell tumor, infants with juvenile granulosa cell tumor of the testis usually present within the first 6 months of life with a painless testicular mass and no associated hormonal symptoms. Normal AFP for age may preoperatively distinguish juvenile granulosa cell tumors from yolk sac tumors, although the wide range of normal AFP values in infancy may limit the utility of results. Infantile juvenile granulosa cell tumors of the testis have been shown to behave in a benign fashion with no evidence for recurrence or metastases (Harms and Kock 1997).

Leydig cell tumors generally occur in children 5–10 years of age or in older men, and may produce testosterone or estrogen. Increase in estrogen may be mediated by direct production of estrogen by the tumor cells or by aromatization of testosterone. Leydig cell tumors may elaborate 17-ketosteroids and thus may present with precocious puberty and/or gynecomastia.

7.2.4 Genetics

Testicular juvenile granulosa cell tumors may be seen in normal boys or in those with abnormalities of the Y chromosome or ambiguous genitalia, thus chromosomal assessment may be informative. As with females, testicular Sertoli cell lesions may be noted in males with PJS. Testicular sex cord-stromal tumors with annular tubules are more likely to show progesterone production and to be bilateral in the setting of PJS. Carney syndrome may be associated with bilateral large cell calcifying Sertoli cell tumors or with Leydig cell tumors (Brown et al. 2007). Testicular tumors in patients with genetic syndromes may be indolent (Kratzer et al. 1997).

A distinctive testicular tumor observed in boys with PJS is multifocal intratubular neoplasia of large Sertoli cells. These tumors may be associated with estradiol production and advanced bone age and generally follow an indolent course. Observation may be considered, although orchiectomy may ultimately be required if invasive features develop or to limit hormone production (Ulbright et al. 2007).

Table 7.2 Children's Oncology Group staging criteria for prepubertal testis germ cell tumors

Stage	Extent of disease
I	Limited to testis (testes), completely resected by high inguinal orchiectomy; no clinical, radiographic or histologic evidence of disease beyond the testes. Patients with normal or unknown tumor markers at diagnosis must have a negative ipsilateral retroperitoneal node sampling to confirm stage I disease if radiographic studies demonstrate lymph nodes >2 cm. Patients who have undergone scrotal orchiectomy with high litigation of cord are stage I
II	Transscrotal biopsy; microscopic disease in scrotum or high in spermatic cord (<5 cm from proximal end). Failure of tumor markers to normalize or decrease with an appropriate half-life
III	Retroperitoneal lymph node involvement, but no visceral or extra-abdominal involvement. Lymph nodes >4 cm by CT or >2 cm and <4 cm with biopsy proof
IV	Distant metastases, including liver

From Children's Oncology Group germ cell tumor protocol. Used with permission

7.2.5 Clinical Evaluation

Clinical evaluation of patients suspected to have a stromal testicular tumor should include a thorough history and physical examination with careful attention to family history and signs of any tumor predisposition syndrome. On clinical examination, careful attention to pubertal status is critical; these observations may guide management decisions. For patients in the peripubertal period, histologic assessment of a small rim of normal testis tissue may guide surgical management and determine the need for staging procedures (see Table 7.2).

7.2.6 Pathology

Inhibin A and B have both been shown to be elevated in some children with testicular sex cord-stromal tumor. However, the diagnostic utility of such laboratory values is limited by the broad normal range in prepubertal children.

Additionally, some cases of Leydig cell tumors express estrogen or progesterone receptors.

7.2.7 Principles of Surgical Therapy

Testis-sparing surgery is considered reasonable for those with juvenile granulosa cell tumors if sufficient normal testicular tissue is present (Shukla et al. 2004). However, the postoperative course may be complicated by scrotal hypotrophy, and its therapeutic advantage may be limited compared to unilateral orchiectomy.

Management of other testis stromal tumors must account for the pubertal status of the patient as well as the histopathologic diagnosis. Sertoli cell tumors, for example, generally behave in a benign fashion in children younger than 5 years of age, thus orchiectomy is considered adequate treatment. The large cell calcifying tumor seen in Carney syndrome is also unlikely to metastasize, and treatment with orchiectomy is usually sufficient (Ross and Kay 2004).

In older children stromal tumors may behave in a more malignant fashion. Children older than 5 years diagnosed with Sertoli cell tumors may benefit from a metastatic evaluation including computed tomography of the chest, abdomen, and pelvis. If found, metastases require aggressive chemotherapy treatment. Histopathology may guide clinical management; large numbers of mitotic figures, local invasion, or poor differentiation may herald a poorer prognosis. A full metastatic evaluation and careful follow-up should be undertaken. The role of adjuvant therapy in patients with concerning histologic features is unclear (Kaplan et al. 1986), as there is scant literature to guide therapeutic decisions.

Leydig cell tumors carry a favorable prognosis in prepubertal boys; orchiectomy or testis-sparing surgery is sufficient treatment for such tumors. In postpubertal males, however, approximately 10 % of Leydig tumors behave in a malignant fashion, with potential metastatic sites of liver, lung, and bone. These tumors have been described as poorly responsive to chemotherapy and radiation, with a reported survival time of 2 months to 17 years (median = 2 years) (Bertram et al. 1991; Al-Agha and Axiotis 2007).

Unspecified stromal tumors may behave in a malignant fashion; pre- and postpubertal boys with such tumors should undergo a full metastatic evaluation (Thomas et al. 2001).

The Prepubertal Testis Tumor Registry has reported 43 patients with stromal testis tumors. Of these, 10 had an unspecified stromal tumor and one developed a metastasis. The single patient with metastatic mixed/undifferentiated stromal tumor died. None of the remaining 32 patients with Sertoli-Leydig or juvenile granulosa cell tumors had metastases at diagnosis. The German MAKEI database includes 42 patients; all presented with localized disease and remain in complete remission after surgical resection (Dominik T. Schneider, 2011, personal communication).

7.2.8 Adult Sex Cord-Stromal Tumors of the Testis

Sex cord-stromal tumors account for <5 % of adult testicular tumors. Some series have noted a generally benign clinical course (Featherstone et al. 2009), but one report describes 8/17 patients with > stage I disease (stages IIA–IIIA). Of these, six eventually died of metastatic disease despite adjuvant therapy (Mosharafa et al. 2003). Thus, although testicular sex cord-stromal tumors are rare in adults, aggressive behaviors may occur, and retroperitoneal lymph node dissection should be considered (Acar et al. 2009).

7.2.9 Adjuvant Therapy

Generally speaking, testicular stromal tumors in stage Ia or Ib should be followed expectantly. Testicular stromal tumors with confirmed evidence of metastatic disease require adjuvant therapy, although little evidence is available to guide recommendations for chemotherapeutic regimens. Based on the authors' experience primarily with ovarian sex cord-stromal tumors, a platinum-based regimen such as cisplatin, etoposide, and bleomycin or cisplatin, etoposide, and ifosfamide is reasonable.

Overall, most data on the clinical behavior and management of stromal tumors of the testis are based on small case series or, in the case of the Prepubertal Testis Tumor Registry, on data obtained at a single time point with limited follow-up (Ross et al. 2002). Further evidence regarding clinical behavior, optimal treatment, and follow-up regimens is needed.

7.2.10 Follow-Up

In tumors with a predisposition for clinically malignant behavior, serial imaging and monitoring of tumor markers is suggested. Long-term follow-up of boys with juvenile granulosa cell tumor diagnosed in infancy is generally not required, but if AFP is elevated at diagnosis, confirmation of an age-appropriate decline in AFP is suggested.

7.3   Summary and Future Research

Sex cord-stromal tumors of the ovary and testis are extremely rare and have varying patterns of clinical behavior. Although low-stage, well-differentiated tumors are often associated with a good prognosis, outcomes are still poor for children with higher stage or less differentiated tumors. Additionally, underlying genetic syndromes may be present in some patients, although their prevalence has not been determined.

The US-based International Ovarian and Testicular Stromal Tumor Registry (OTST Registry) has been established to advance knowledge about the behavior of these tumors, determine pathophysiology and treatment efficacy, and establish comprehensive follow-up guidelines. All patients with ovarian or testicular sex cord-stromal tumors are eligible for enrollment. Central pathology review of available samples is provided as part of Registry participation. For further information, please contact the Registry (www.OTSTregistry.org).

In Europe, patients with ovarian and testicular sex cord-stromal tumors are registered within national germ cell tumor trials or within national rare tumor working groups. The largest worldwide cohort has been collected within the German MAKEI trials and includes more than 200 patients. This registry is also open to international patients. Moreover, there is a growing international network of the European rare tumor groups who have founded the European Cooperative Study Group on Pediatric Rare Tumors (EXPeRT). The EXPeRT group supports international joint analyses on specific rare tumor entities including rare gonadal tumors. Further details are extensively discussed in Rare Tumors in Children and Adolescents (Bisogno 2012).

With data collection efforts expanding for these rare tumors of childhood, much-needed information about genetic predisposition, tumor histopathology, clinical behavior, treatment efficacy, outcomes, and late effects will become available to better serve these children.

References

Acar C, Gurocak S, Sozen S (2009) Current treatment of testicular sex cord-stromal tumors: critical review. Urology 73:1165–1171

Al-Agha OM, Axiotis CA (2007) An in-depth look at Leydig cell tumor of the testis. Arch Pathol Lab Med 131:311–317

Ball A, Wenning J, Van Eyk N (2011) Ovarian fibromas in pediatric patients with basal cell nevus (Gorlin) syndrome. J Pediatr Adolesc Gynecol 24:e5–e7

Beggs AD, Latchford AR, Vasen HF et al (2010) Peutz-Jeghers syndrome: a systematic review and recommendations for management. Gut 59:975–986

Bertram KA, Bratloff B, Hodges GF et al (1991) Treatment of malignant Leydig cell tumor. Cancer 68:2324–2329

Bisogno G (2012) The EXPeRT initiative. In: Schneider DT (ed) Rare tumors in children and adolescents. Springer, Berlin/Heidelberg, pp 327–402

Breen JL, Maxson WS (1977) Ovarian tumors in children and adolescents. Clin Obstet Gynecol 20:607–623

Brown B, Ram A, Clayton P et al (2007) Conservative management of bilateral sertoli cell tumors of the testicle in association with the Carney complex: a case report. J Pediatr Surg 42:E13–E15

Brown J, Shvartsman HS, Deavers MT et al (2005) The activity of taxanes compared with bleomycin, etoposide, and cisplatin in the treatment of sex cord-stromal ovarian tumors. Gynecol Oncol 97:489–496

Brown J, Sood AK, Deavers MT et al (2009) Patterns of metastasis in sex cord-stromal tumors of the ovary: Can routine staging lymphadenectomy be omitted? Gynecol Oncol 113:86–90

Costa MJ, Morris R, Sasano H (1994) Sex steroid biosynthesis enzymes in ovarian sex-cord stromal tumors. Int J Gynecol Pathol 13:109–119

Featherstone JM, Fernando HS, Theaker JM et al (2009) Sex cord stromal testicular tumors: a clinical series–uniformly stage I disease. J Urol 181:2090–2096; discussion 2096

Fefferman NR, Pinkney LP, Rivera R et al (2003) Sclerosing stromal tumor of the ovary in a premenarchal female. Pediatr Radiol 33:56–58

Gershenson DM (1994) Management of early ovarian cancer: germ cell and sex cord-stromal tumors. Gynecol Oncol 55:S62–S72

Gershenson DM, Morris M, Burke TW et al (1996) Treatment of poor-prognosis sex cord-stromal tumors of the ovary with the combination of bleomycin, etoposide, and cisplatin. Obstet Gynecol 87:527–531

Gibbon DG (2005) Conservative management of sex cord tumors with annular tubules of the ovary in women with Peutz-Jeghers syndrome. J Pediatr Hematol Oncol 27:630–632

Harms D, Kock LR (1997) Testicular juvenile granulosa cell and Sertoli cell tumours: a clinicopathological study of 29 cases from the Kiel Paediatric Tumour Registry. Virchows Arch 430:301–309

Hassan E, Creatsas G, Deligeorolgou E et al (1999) Ovarian tumors during childhood and adolescence. A clinicopathological study. Eur J Gynaecol Oncol 20:124–126

Heravi-Moussavi A, Anglesio MS, Cheng SW et al (2012) Recurrent somatic DICER1 mutations in nonepithelial ovarian cancers. N Engl J Med 355:234–242

Hill DA, Ivanovich J, Priest JR et al (2009) DICER1 mutations in familial pleuropulmonary blastoma. Science 325:965

Irving JA, Young RH (2009) Microcystic stromal tumor of the ovary: report of 16 cases of a hitherto uncharacterized distinctive ovarian neoplasm. Am J Surg Pathol 33:367–375

Jamieson S, Butzow R, Andersson N et al (2010) The FOXL2 C134W mutation is characteristic of adult granulosa cell tumors of the ovary. Mod Pathol 23:1477–1485

Kaplan GW, Cromie WJ, Kelalis PP et al (1986) Gonadal stromal tumors: a report of the Prepubertal Testicular Tumor Registry. J Urol 136:300–302

Kato N, Romero M, Catasus L et al (2004) The STK11/LKB1 Peutz-Jegher gene is not involved in the pathogenesis of sporadic sex cord-stromal tumors, although loss of heterozygosity at 19p13.3 indicates other gene alteration in these tumors. Hum Pathol 35:1101–1104

Kim GJ, Georg I, Scherthan H et al (2010a) Dicer is required for Sertoli cell function and survival. Int J Dev Biol 54:867–875

Kim T, Sung CO, Song SY et al (2010b) FOXL2 mutation in granulosa-cell tumours of the ovary. Histopathology 56:408–410

Koonings PP, Campbell K, Mishell DR Jr et al (1989) Relative frequency of primary ovarian neoplasms: a 10-year review. Obstet Gynecol 74:921–926

Kratzer SS, Ulbright TM, Talerman A et al (1997) Large cell calcifying Sertoli cell tumor of the testis: contrasting features of six malignant and six benign tumors and a review of the literature. Am J Surg Pathol 21:1271–1280

Merras-Salmio L, Vettenranta K, Mottonen M et al (2002) Ovarian granulosa cell tumors in childhood. Pediatr Hematol Oncol 19:145–156

Mosharafa AA, Foster RS, Bihrle R et al (2003) Does retroperitoneal lymph node dissection have a curative role for patients with sex cord-stromal testicular tumors? Cancer 98:753–757

Ozulker T, Ozpacaci T, Ozulker F et al (2010) Incidental detection of Sertoli-Leydig cell tumor by FDG PET/CT imaging in a patient with androgen insensitivity syndrome. Ann Nucl Med 24:35–39

Ramirez PT, Schmeler KM, Milam MR et al (2008) Efficacy of letrozole in the treatment of recurrent platinum- and taxane-resistant high-grade cancer of the ovary or peritoneum. Gynecol Oncol 110:56–59

Rio Frio T, Bahubeshi A, Kanellopoulou C et al (2011) DICER1 mutations in familial multinodular goiter with and without ovarian Sertoli-Leydig cell tumors. JAMA 305:68–77

Rishi M, Howard LN, Bratthauer GL et al (1997) Use of monoclonal antibody against human inhibin as a marker for sex cord-stromal tumors of the ovary. Am J Surg Pathol 21:583–589

Ross JH, Kay R (2004) Prepubertal testis tumors. Rev Urol 6:11–18

Ross JH, Rybicki L, Kay R (2002) Clinical behavior and a contemporary management algorithm for prepubertal testis tumors: a summary of the Prepubertal Testis Tumor Registry. J Urol 168:1675–1678; discussion 1678–1679

Rutgers JK (2011) The case reported as bilateral Sertoli-Leydig cell tumors in a 61-year-old woman with uterine aplasia may instead represent complete androgen insensitivity syndrome. Int J Gynecol Pathol 30:395

Satoh M (1991) Histogenesis and organogenesis of the gonad in human embryos. J Anat 177:85–107

Schneider DT, Calaminus G, Wessalowski R et al (2002) Therapy of advanced ovarian juvenile granulosa cell tumors. Klin Padiatr 214:173–178

Schneider DT, Calaminus G, Wessalowski R et al (2003a) Ovarian sex cord-stromal tumors in children and adolescents. J Clin Oncol 21:2357–2363

Schneider DT, Janig U, Calaminus G et al (2003b) Ovarian sex cord-stromal tumors–a clinicopathological study of 72 cases from the Kiel Pediatric Tumor Registry. Virchows Arch 443:549–560

Schneider DT, Terenziani M, Cecchetto G (2012) Gonadal and extragonadal germ cell tumors, sex cord stromal and rare gonadal tumors. In: Schneider DT (ed) Rare tumors in children and adolescents. Springer, Berlin/Heidelberg, pp 327–402

Schultz KA, Pacheco MC, Yang J et al (2011) Ovarian sex cord-stromal tumors, pleuropulmonary blastoma and DICER1 mutations: a report from the International Pleuropulmonary Blastoma Registry. Gynecol Oncol 122:246–250

Schultz KA, Sencer SF, Messinger Y et al (2005) Pediatric ovarian tumors: a review of 67 cases. Pediatr Blood Cancer 44:167–173

Shah SP, Kobel M, Senz J et al (2009) Mutation of FOXL2 in granulosa-cell tumors of the ovary. N Engl J Med 360:2719–2729

Shukla AR, Huff DS, Canning DA et al (2004) Juvenile granulosa cell tumor of the testis: contemporary clinical management and pathological diagnosis. J Urol 171:1900–1902

Sivathondan Y, Salm R, Hughesdon PE et al (1979) Female adnexal tumour of probable Wolffian origin. J Clin Pathol 32:616–624

Slade I, Bacchelli C, Davies H et al (2011) DICER1 syndrome: clarifying the diagnosis, clinical features and management implications of a pleiotropic tumour predisposition syndrome. J Med Genet 48:273–278

Stacher E, Pristauz G, Scholz HS et al (2010) Bilateral ovarian well-differentiated Sertoli-Leydig cell tumors with heterologous elements associated with unilateral serous cystadenoma–a case report. Int J Gynecol Pathol 29:419–422

Sumi T, Ishiko O, Maeda L et al (2002) Adenocarcinoma arising from respiratory ciliated epithelium in mature ovarian cystic teratoma. Arch Gynecol Obstet 267:107–109

Tanaka Y, Sasaki Y, Nishihira H et al (1992) Ovarian juvenile granulosa cell tumor associated with Maffucci's syndrome. Am J Clin Pathol 97:523–527

Thomas JC, Ross JH, Kay R (2001) Stromal testis tumors in children: a report from the prepubertal testis tumor registry. J Urol 166:2338–2340

Ulbright TM, Amin MB, Young RH (2007) Intratubular large cell hyalinizing sertoli cell neoplasia of the testis: a report of 8 cases of a distinctive lesion of the Peutz-Jeghers syndrome. Am J Surg Pathol 31:827–835

Vaz RM, Turner C (1986) Ollier disease (enchondromatosis) associated with ovarian juvenile granulosa cell tumor and precocious pseudopuberty. J Pediatr 108:945–947

Velasco-Oses A, Alonso-Alvaro A, Blanco-Pozo A et al (1988) Ollier's disease associated with ovarian juvenile granulosa cell tumor. Cancer 62:222–225

White LC, Buchanan KD, O'Leary TD et al (2003) Direct laparoscopic venous sampling to diagnose a small Sertoli-Leydig tumor. Gynecol Oncol 91:254–257

Young RH (2005) Sex cord-stromal tumors of the ovary and testis: their similarities and differences with consideration of selected problems. Mod Pathol 18(Suppl 2):S81–S98

Zheng W, Senturk BZ, Parkash V (2003) Inhibin immunohistochemical staining: a practical approach for the surgical pathologist in the diagnoses of ovarian sex cord-stromal tumors. Adv Anat Pathol 10:27–38

Late Effects in Testicular Cancer Survivors

8

Clair Beard

Contents

C. Beard, MD
Department of Radiation Oncology,
Brigham and Women's Hospital, Boston, MA, USA

Department of Radiation Oncology,
Dana-Farber Cancer Institute, Boston, MA, USA
e-mail: cbeard@lroc.harvard.edu,
clair_beard@dfci.harvard.edu

Treatment options for men with testicular cancer vary by type of tumor (seminomatous vs. nonseminomatous) and stage at presentation. Options range from surveillance, with therapy reserved only for those who relapse, to retroperitoneal lymph-node dissection, external-beam irradiation, cisplatin-based chemotherapy, resection of metastatic lesions, and bone marrow transplantation; see Table 8.1. Late effects appear to be most related to the type of therapy received, with the fewest long-term complications seen after simple orchiectomy and the most after high-dose or salvage chemotherapy or in patients who received both radiotherapy and chemotherapy. Age at diagnosis, comorbidity, and exposure to other toxins, such as tobacco, are also relevant. The study of late effects is important because testicular-cancer survivors who are long since cured of their malignancies have a higher than expected all-cause mortality rate over time due to late effects. Increased risks of contralateral testicular cancer, second non-germ cell tumors, and cardiovascular disease including hypertension, metabolic syndrome, diabetes, and renal dysfunction have all been reported, as have infertility, neurological effects such as neuropathy, Raynaud's and ototoxicity, and late psychosocial effects. This chapter is a summary of what is known about late effects in testicular-cancer survivors as well as areas of ongoing or future research.

A.L. Frazier, J.F. Amatruda (eds.), *Pediatric Germ Cell Tumors*, Pediatric Oncology 1,
DOI 10.1007/978-3-642-38971-9_8, © Springer-Verlag Berlin Heidelberg 2014

Table 8.1 Treatment options for men with testicular cancer

Early stage	Good risk	Intermediate risk	High risk or 2nd-line therapy	Palliative
Seminoma				
Surveillance after orchiectomy:	Radiotherapy:	Chemotherapy:	Chemotherapy:	
6–20 CT scans over 5–10 years	*Historical*: extended field to 35–40 Gy	BEP×4	VeIP	
Radiotherapy:	Mediastinal irradiation	VIP×4	TIP	
Historical: 30 Gy to extended field or "dog-leg" field	Left supraclavicular irradiation		High-dose therapy:	
Modern: 20 Gy to small field "PA-strip"	*Modern*: smaller field to retroperitoneum and common iliac nodes to 26–35 Gy		High-dose Carbo-E with stem-cell support or	
Carboplatin:	Chemotherapy:		PI–CE with stem-cell support	
1 dose AUC7	BEP × 3		Surgical removal of residual masses as indicated	
	EP × 4			
Nonseminoma				
Surveillance after orchiectomy:	Surgery:	Chemotherapy:	Chemotherapy:	Chemotherapy:
Surgery:	Nerve-sparing RPLND	BEP×4	VeIP	GEMOX
Nerve-sparing RPLND*	Chemotherapy:	VIP×4	TIP	GEM-TAXOL
Chemotherapy:	BEP×3	Surgical removal of residual masses as indicated	High-dose therapy:	
BEP×1	EP×4		High-dose Carbo-E with stem-cell support or	
BEP×2			PI–CE with stem-cell support	
			Surgical removal of residual masses as indicated	

Risk categories for metastatic disease defined by the International Germ Cell Cancer Collaborative Group

Standard chemotherapy regimens:	High-dose regimens:
BEP: bleomycin, etoposide, cisplatin	carboplatin, etoposide with stem-cell support
EP: etoposide, cisplatin	*PI–CE*: paclitaxel, ifosfamide followed by carboplatin, etoposide with stem-cell support
VIP: etoposide, ifosfamide, cisplatin	Palliative:
VeIP: vinblastine, ifosfamide, cisplatin	*GEMOX*: gemcitabine, oxaliplatin
TIP: paclitaxel, ifosfamide, cisplatin	*GEM-TAXOL*: gemcitabine, paclitaxel

*Retroperitoneal lymph node dissection

8.1 Second Malignant Neoplasms

Second malignant neoplasms are the greatest threat to longevity for testicular-cancer survivors. An increased risk of solid tumors, leukemias, and contralateral testicular cancer has been reported; see Table 8.2. Reports show associations with radiotherapy, chemotherapy, and combined modality therapy.

Table 8.2 Non-germ cell malignancies in testicular-cancer survivors

Study population	No. of subjects	Years of diagnosis	Follow-up time (years)	Treatment	Tumors observed	RR	95 % CI
All malignancies							
14 population-based tumor registries in Europe and North America (Travis et al. 2005)[a,b,c,d]	40,576	1943–2001	Mean=11.3	Any	1,694[e,f]	1.9	1.8–2.1
				RT	892	2	1.9–2.2
				CT	35	1.8	1.3–2.5
				RT+CT	25	2.9	1.9–4.2
13 international population-based cancer registries (Richiardi et al. 2007)[d,g]	29,511	1943–2000	Median=8.3	Any	1,811[h]	1.7	1.6–1.7
Swedish Family-Cancer Database (Hemminki et al. 2010)	5,533	1980–2000	NA	Any	274[i]	2	1.8–2.2
Netherlands testicular-cancer survivors cohort (van den Belt-Dusebout et al. 2007)	2,707	1965–1995	Median=17.6	Any	270[j]	1.7	1.5–1.9
				Surgery	12	0.7	0.4–1.3
				RT	199	1.7	1.5–2.0
				CT	23	1.4	0.9–2.1
				RT+CT	29	3	2.0–4.4
				SDRT[k]	NA	2.6	1.7–4.0
				SDRT (26–35 Gy)[k]	NA	2.3	1.5–3.6
				SDRT (40–50 Gy)[k]	NA	3.2	2.1–5.1
				SDRT+MRT[k]	NA	3.6	2.1–6.0
				PVB/PEB only[k]	NA	2.1	1.3–2.4
Norwegian Radium Hospital cohort (Wanderas et al. 1997)	2,006	1952–1990	Mean=12.5	Any	153[l]	1.7	1.4–1.9
				RT	130	1.6	1.3–1.9
				CT	4	1.3	0.4–3.4
				RT+CT	15	3.5	2.0–5.8

(continued)

Table 8.2 (continued)

Study population	No. of subjects	Years of diagnosis	Follow-up time (years)	Treatment	Tumors observed	RR	95 % CI
Leukemia specific							
Case–control study using 8 population-based cancer registries (Travis et al. 2000)[m]	18,567	1970–1993		No RT or CT	4	1[n]	(0.7–22)
				RT	22	3.1	(1.0–40)
				CT	8	5	(0.5–58)
				RT+CT	2	5.1	

Reprinted with permission: Travis et al. (2011)

Abbreviations: RR relative risk, 95 % CI 95 % confidence interval, *Any* any singular or combination of surgery, radiation, and chemotherapy, *RT* radiation therapy, *CT* chemotherapy, *NA* not available, *SDRT* subdiaphragmatic radiation therapy, *MRT* mediastinal radiation therapy, *PVB* cisplatin, vinblastine, bleomycin, *PEB* cisplatin, etoposide, bleomycin

[a]Excludes contralateral testicular cancers

[b]Registries and periods of treatment: Canada (Ontario, 1964–2000), Denmark (1943–1998), Finland (1953–1998), Norway (1953–1999), Sweden (1958–2001), and nine registries in the United States (1973–1999) that participate in the National Cancer Institute's Surveillance, Epidemiology, and End Results (SEER) Program, including the states of Connecticut, Hawaii, Iowa, New Mexico, and Utah, as well as the metropolitan areas of San Francisco–Oakland, Detroit, Seattle–Puget Sound (from 1974), and Atlanta (from 1975)

[c]A subset of patients used in the study by Wanderas et al. (1997) are included in this study with longer follow-up

[d]4 cancer registries were used by both Travis et al. (2005) and Richiardi et al. (2006): Denmark, Finland, Norway, and Sweden

[e]Number of tumors seen in patients who had at least 10 years of follow-up

[f]Only solid tumors recorded in this study

[g]Registries and periods of treatment: New South Wales, Australia (1972–1997); British Colombia, Canada (1970–1998); Manitoba, Canada (1970–1998); Saskatchewan, Canada (1967–1998); Denmark (1943–1997); Finland (1953–1998); Iceland (1955–2000); Norway (1953–1999); Singapore (1968–1992); Slovenia (1961–1998); Zaragoza, Spain (1978–1998); Sweden (1961–1998); Scotland, United Kingdom (1975–1996)

[h]38 cases of AML, RR=3.6 (95 % CI, 2.6–5.0); 13 cases ALL, RR=1.0 (95 % CI, 0.6–1.7); 23 cases "other leukemia" RR=3.4 (95 % CI, 2.2–5.2)

[i]18 cases of leukemia total, RR=3.9 (95 % CI, 2.3–6.2); 5 AML, RR=5.6 (95 % CI, 1.6–13); 2 ALL, RR=6.5 (95 % CI, 0.8–23); 4 CML, RR=6.7 (95 % CI, 1.8–17); 3 CLL, RR=2.0 (95 % CI, 0.4–5.8); 4 "other leukemia," RR=3.8 (95 % CI, 1.0–9.8)

[j]6 cases of leukemia, RR=1.6 (95 % CI, 0.6–3.5)

[k]Findings part of a multivariate analysis of 263 patients who had both a second malignancy and a cardiovascular event; adjusted for smoking and age at diagnosis

[l]6 cases of leukemia, RR=1.9 (95 % CI, 0.7–4.1)

[m]Registries used: Iowa, Connecticut, New Jersey, Ontario, Denmark, the Netherlands, Sweden, and Finland

[n]Surgery-only group was the referent group

8.1.1   Solid Tumors

The largest study to date reviewed the risk of invasive solid cancers for 40,576 men with a history of testicular cancer (Travis et al. 2005). The testicular cancers were all diagnosed between 1943 and 2001 and therefore encompassed multiple historical treatment regimens, including the shift to lower radiation doses and the use of cisplatin-based chemotherapy regimens, either with or without radiotherapy. The study investigated the risk of a second solid tumor over the man's lifetime; leukemia and contralateral testicular cancer were not included in the analysis nor were tertiary or higher-order tumors, meaning that the total number of subsequent cancers may be underestimated by the data. A total of 20,984, 7,885, and 2,065 patients had follow-up periods of at least 10, 20, and 30 years, respectively. For patients who survived at least 10 years after diagnosis, the relative risk of developing a solid tumor compared to the general population was calculated for each treatment group. Patients who received radiotherapy only had a relative risk (RR) of 2.0 (95 % confidence interval [CI], 1.9–2.2). For those who received chemotherapy alone, the RR was 1.8 (95 % CI, 1.3–2.5). Patients who received both radiotherapy and chemotherapy had a RR of 2.9 (95 % CI, 1.9–4.2). The authors note that although the RR for patients who received combined modalities was higher than that for patients treated with either modality alone, the differences between the three groups were not statistically significant. Of the solid tumors reported to have an increased incidence, lung (RR = 1.5; 95 % CI, 1.2–1.7), colon (RR = 2.0; 95 % CI, 1.7–2.5), bladder (RR = 2.7; 95 % CI, 2.2–3.1), pancreas (RR = 3.6; 95 % CI, 2.8–4.6), and stomach (RR = 4.0; 95 % CI, 3.2–4.8) make up almost 60 % of the total excess. A patient's cumulative risk of developing a solid tumor with 40 years of follow-up would be 36 % for men with seminomas and 31 % for those with nonseminomatous germ cell tumors compared to 23 % for the general population. The risk of developing a solid tumor increased over time and persisted for at least 30 years after treatment. Using the trends identified by this study, the authors postulated that

a 20-year-old patient diagnosed with seminoma would have a cumulative risk of developing a solid cancer by age 75 of 47 %; the equivalent risk for patients diagnosed at ages 35 and 50 would be 36 % and 28 %, respectively. Standard radiotherapy fields with organs at risk are shown in Fig. 8.1.

Other supporting data are provided by Wanderas et al. (1997) who observed a relative risk of 1.65 (95 % CI, 1.4–1.9) for any secondary non-germ cell malignancy in a cohort of 2006 Norwegian patients treated between 1952 and 1990. A total of 153 tumors were observed, of which 147 were solid tumors; the largest proportions were gastrointestinal (n = 39, RR = 1.81; 95 % CI, 1.3–2.5) and urogenital (n = 29, RR = 1.36; 95 % CI, 0.9–1.8). As in the study above, patients who received both radiotherapy and chemotherapy had the highest RR at 3.54 (95 % CI, 2.0–5.8) with radiotherapy alone having the next highest risk (RR = 1.58; 95 % CI, 1.3–1.9). No increase in risk was seen in the group that received only chemotherapy.

Although an increased risk of solid tumors for patients who received chemotherapy alone was not seen in older smaller studies (Pedersen-Bjergaard et al. 1991; Bokemeyer and Schmoll 1993; Kollmannsberger et al. 1999), the larger 2005 study (Travis et al. 2005) did show an overall increase. This was also noted in a subsequent 2007 study that reviewed the treatment-specific risks of second malignancies and cardiovascular disease in 5-year survivors of testicular cancer (van den Belt-Dusebout et al. 2007). This tumor registry study looked at a cohort of 2,707 patients with testicular cancer from the Netherlands who were at least 5 years beyond completion of therapy. Patients were treated between 1965 and 1995 with surgery followed either by surveillance or by treatment with adjuvant radiotherapy, chemotherapy, or both. After a median follow-up period of 17.6 years, there was no absolute increase in the standardized incidence rate (SIR) for second malignancies in general; however, a multivariate Cox regression analysis of risk factors for development of second malignancies showed an increased hazard ratio (HR) of 2.1 (95 % CI, 1.4–3.1) for patients treated with either cisplatin, vinblastine, and bleomycin or bleomycin, etoposide, and

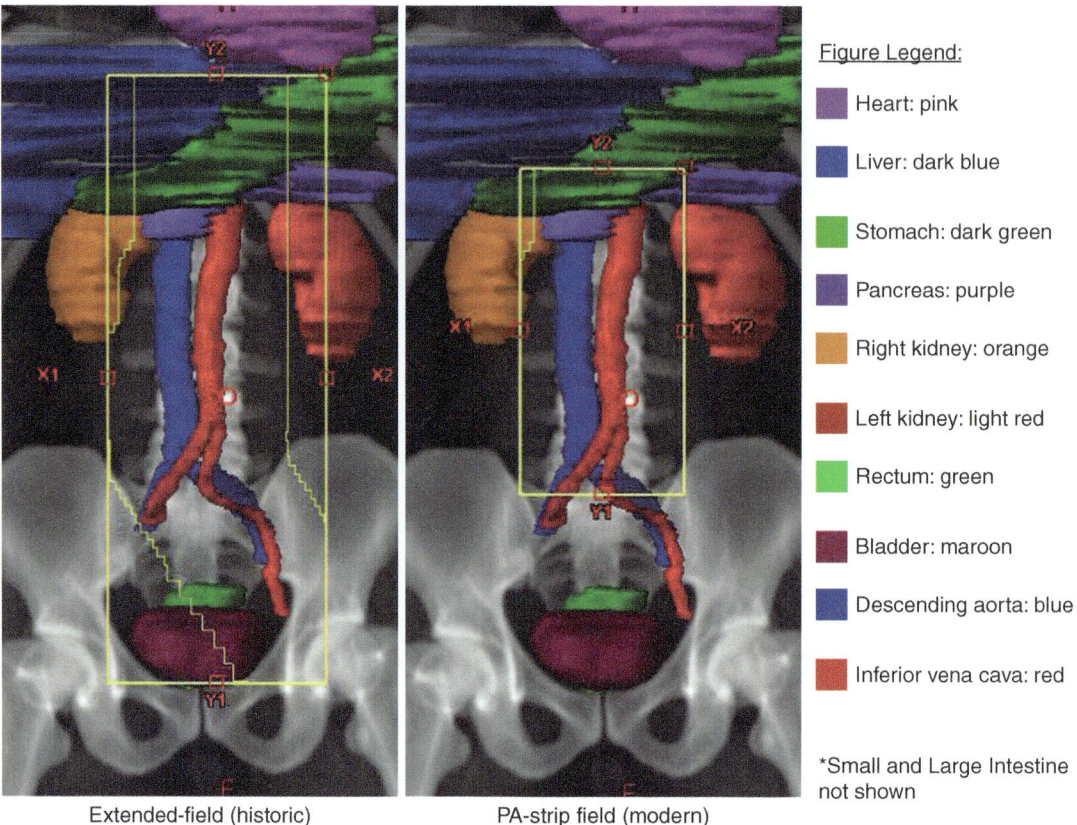

Extended-field (historic)   PA-strip field (modern)

Fig. 8.1 Radiotherapy fields and organs at risk (small and large intestine not shown)

cisplatin without radiation. It is unclear why newer studies show an association with chemotherapy; however, two issues may be at play. First, the large number of patients in the 2005 study may allow more subtle effects to become apparent. Second, the risk of a second solid tumor appears to increase throughout a patient's life. The newer studies showing an increased long-term risk associated with chemotherapy may reflect longer follow-up for patients in the era of increased use of cisplatin, which was introduced into the clinic in 1978 and widely used by the 1980s. Cisplatin substantially increases the survival time of patients with metastatic testicular cancer, allowing more patients to survive long enough to develop late effects.

8.1.2 Hematologic Malignancies

An increased incidence of leukemia after treatment for testicular cancer has also been shown,

however, not to the same extent as solid tumors. To look more closely at the specific risk of leukemia, Travis et al. performed a case–control study of the prevalence of leukemia and its association with the use of radiotherapy and chemotherapy for testicular cancer, using a cohort of 18,567 men treated between 1970 and 1993 (Travis et al. 1996). The cases were 36 men with testicular cancer and secondary leukemia; they were matched with 106 controls of the same age, year of testicular-cancer diagnosis, and no secondary malignancy up to the time that the case developed leukemia. Unlike with solid tumors, the median time from diagnosis of testicular cancer to diagnosis of leukemia was 5 years, with a mean of 6.8 years; 25 % of the leukemias were diagnosed after 10 years. The study observed no overall increase in leukemia rate for the entire cohort of patients who received only radiation (RR=3.1; 95 % CI, 0.7–22); however, radiation doses ≥20 Gy were associated with a significant increase (RR = 19.7; 95 % CI, 1.5–590). Patients treated with only pelvic and abdominal field radiation had a RR of 2.9 (95 % CI, 0.6–21) whereas those with chest irradiation as well had a RR of 11.2 (1.5–123). Similar to the radiotherapy effect, patients who received any cisplatin-based regimens, with or without etoposide or chlorambucil, did not have a significant increase in their risk of leukemia (RR = 5.0; 95 % CI, 1.0–40); however, when cumulative cisplatin dose was viewed as a continuous variable, a dose of 650 mg conferred a RR of 3.2 (95 % CI, 1.5–8.4) and RR increased with increasing amounts of cisplatin (e.g., for 1,000 mg, the RR was 5.9 [95 % CI, 2.0–26]). When taking into account the use of cisplatin and radiation, no effect on risk could be detected for etoposide or bleomycin.

8.1.3 Contralateral Testicular Cancer

Another concern for men with testicular cancer is the possibility of contralateral testicular cancer (CLTC) although the overall risks are low. A population-based study of 29,515 men from the United States diagnosed between 1973 and 2001 showed a 15-year cumulative risk of 1.9 % (95 % CI, 1.7–13.9) for development of a CLTC at least 2 months after diagnosis of the first testicular cancer (Fossa et al. 2005). This was a bit lower than the risk observed in other studies, including ones from the Netherlands (15-year cumulative risk 2.4 %; 95 % CI, 1.4–3.9) (van Leeuwen et al. 1993), Denmark (20-year cumulative risk 5.2 %; 95 % CI, 3.7–6.7) (Osterlind et al. 1991), and Norway (15-year cumulative risk 3.9 %; 95 % CI, 2.8–5) (Wanderas et al. 1997). In the US study, univariate analysis showed nonseminoma histology, treatment with chemotherapy, and distant metastatic disease at diagnosis all decreased the risk of CLTC. However, on multivariate analysis, only nonseminoma histology retained its significance, with a HR of 0.6 (95 % CI, 0.46–0.79). Patients with nonseminomatous tumors were much more likely to receive chemotherapy; only one of the 173 men who initially had seminoma was treated with chemotherapy. In looking at age and histol-

ogy, men 30 years old or younger at initial diagnosis who had a seminomatous testicular cancer had the highest risk (15-year cumulative risk = 3.1 %; 95 % CI, 2.4–4.0), whereas the risk for men diagnosed over the age of 30 with a non-seminomatous testicular cancer was 1.2 % (95 % CI, 0.8–1.8). The observed-over-expected ratio of any contralateral testicular tumor was 12.4 (95 % CI, 11.2–13.9). The median latency was 63 months with a range of 3–223 months.

8.2 Cardiovascular Disease

The second most important long-term effect of therapy for germ cell tumors is cardiovascular disease. Published information is available pertaining to long-term survivors of testicular cancer, ovarian germ cell tumors, and pediatric germ cell tumors although, as with second malignancies, most of the available data come from the testicular-cancer cohorts. One important note that becomes apparent upon reviewing the literature is that the definition of cardiovascular disease differs among studies based on the study parameters.

An increase in cardiovascular risk factors and disease in long-term survivors of testicular cancer has been well documented; see Table 8.3 (Huddart et al. 2003; Zagars et al. 2004; van den Belt-Dusebout et al. 2006, 2007; Wethal et al. 2007; Haugnes et al. 2010). A robust study from the Netherlands reviewed the long-term risk of cardiovascular disease in 5-year testicular-cancer survivors (van den Belt-Dusebout et al. 2006). In a cohort of 2,512 men treated between 1965 and 1995, that study found 694 events in 434 patients. Acute myocardial infarction (MI) ($n = 141$) and angina pectoris ($n = 150$) were the most common. The overall standardized incidence ratio for MI and angina pectoris combined was 1.17 (95 % CI, 1.04–1.31). Interestingly, the nonseminoma patients with an attained age of either <45 years or 45–54 years had a higher risk of MI than the patients 55 or older. Using Cox regression analysis, an increased risk of cardiovascular disease was seen with the use of mediastinal radiation (RR = 3.0); the combination of cisplatin, vinblastine, and bleomycin (PVB) (RR = 1.9); and a

Table 8.3 Cardiovascular disease in testicular-cancer survivors

Study population	No. of subjects	Years of diagnosis	Follow-up time (years)	Treatment	Events observed	RR	95 % CI
Netherlands testicular-cancer survivors cohort (van den Belt-Dusebout et al. 2006)[a,b]	2,339	1965–1995	Median = 18.4	Any	291[c]	1.2	1.0–1.3
				Surgery	25	0.9	0.6–1.4
				RT	183	1.1	0.9–1.2
				IDRT	143	0.9	0.8–1.1
				SDRT	40	2.5	1.8–3.4
				CT	42	1.4	1.0–1.8
				CT+SRT	18	1.8	1.1–2.9
				CT+MRT	17	3.0	1.7–4.8
				Unknown	6	1.5	0.5–3.2
Royal Marsden National Health Service Trust (Huddart 2003)[d,e]	992	1982–1992	Median = 10.2	Any	68[f]	1[g]	
				Surgery	9	2.6	1.2–5.8
				CT	26	2.4	1.0–5.5
				RT	22	2.8	1.1–7.1
				RT+CT	11		
Norwegian University Hospitals (Haugnes 2010)[h,i]	990	1980–1994	Median = 19	Any	74[j]	1[g]	
				Surgery	7	2.3	1.0–5.3
				RT	34	4.7	1.6–14.1
				CT+RT	6	2.6	1.1–5.9
				CT	27	1.9	0.7–5.0
				CVB	NA	4.7	1.8–12.2
				PEB	NA		

Abbreviations: *RR* relative risk, *95 % CI*, 95 % confidence interval, *Any* any singular or combination of surgical, radiation, and chemotherapy, *RT* radiation therapy, *IDRT* infra-diaphragmatic radiation, *SDRT* supradiaphragmatic radiation, *CT* chemotherapy, *SDRT* subdiaphragmatic radiation therapy, *MRT* mediastinal radiation therapy, *PVB* cisplatin, vinblastine, bleomycin, *PEB* cisplatin, etoposide, bleomycin

aEvent defined as acute myocardial infarction and angina pectoris

bEvents occurring <5 years from diagnosis were excluded

c141 myocardial infarctions; 150 angina pectoris

dEvent defined as acute myocardial infarction, angina, and cardiac abnormality listed on assessment form at long-term follow-up visit or cardiac surgery for coronary artery disease

eFindings extracted from a multivariate logistic regression analysis adjusted for age at presentation

f41 angina or chest pain, 9 myocardial infarctions, and18 sudden or cardiac deaths

gReferent group

hEvent defined as myocardial infarction, angina, stroke, transient ischemic attack, carotid stenosis, aneurysm of aorta or renal artery, or intermittent claudication

iFindings extracted from an age-adjusted Cox regression analysis

jIncludes 29 myocardial infarctions, 37 angina, and 12 stroke recent history of smoking (RR = 1.8). Mediastinal radiation was associated with a RR of 3.7 for an MI and 3.1 for heart failure, and PVB was associated with an increased risk of MI (RR = 1.9) and peripheral vascular disease (RR = 2.2). The same group took a slightly larger cohort and looked at the treatment-specific risks of cardiovascular disease (van den Belt-Dusebout et al. 2007). Subdiaphragmatic radiation alone was not associated with an increased risk of cardiovascular disease; however, men treated with subdiaphragmatic and mediastinal radiation had a HR of 3.0. Patients who received only chemotherapy had an increased HR of 1.7; however, the authors did not differentiate between cisplatin, vinblastine, and bleomycin and cisplatin, etoposide, and bleomycin. This increase was similar to that seen in patients who smoked (HR = 1.8).

Another study reviewed data on 992 patients treated between 1982 and 1992 for testicular cancer with a median follow-up time of 10.2 years (Huddart et al. 2005). Cardiac events, defined as MI, angina or chest pain, and sudden or cardiac deaths, as well as risk factors for cardiac disease including smoking, cholesterol levels, hypertension, body mass index, and biochemical markers for renal injury or hormonal imbalance were noted. The patients who received neither radiation nor chemotherapy were used as the referent group. Radiation was primarily given to a dose of 30 Gy using the extended field "dog-leg" configuration; however, 8 % of patients also received mediastinal radiation. Two thirds of the patients treated with chemotherapy received bleomycin, etoposide, and cisplatin (BEP), and the other third received carboplatinum, etoposide, and bleomycin (JEB). Patients who received both chemotherapy and radiation were analyzed as a separate group. Overall, 68 patients (6.8 %) had cardiac events, including 9 with an MI and 18 who had cardiac or sudden death. In the multivariate age-adjusted logistic regression analysis, radiation therapy was associated with an increased RR for cardiac events; RR was 2.40 (95 % CI, 1.04–5.45) for patients treated with radiation only and was 2.78 (95 % CI, 1.09–7.07) for patients who received radiation in combination with chemotherapy. The risk remained when

patients who received mediastinal radiation were excluded (unlike the results of the Dutch study discussed above). It should be noted, however, that the field arrangements used for patients who did not undergo mediastinal radiotherapy extended to the top of vertebral body T-11; this could have allowed for direct ventricular irradiation and may have contributed to the relatively high RR (Beard C, 2011, personal communication). The chemotherapy-only group did not have a significantly increased RR on univariate analysis; however, in an age-adjusted logistic regression model, their RR was significant at 2.59 (95 % CI, 1.15–5.84). Although multiple variables were identified as being associated with the development of cardiovascular disease, including higher age at diagnosis, free thyroxine level, sodium protein, albumin, and magnesium levels, as well as radiotherapy, only increasing age ($p = 0.002$), magnesium ($p = 0.021$), free thyroxine ($p - 0.033$), and total protein levels ($p = 0.030$) remained significant in the multivariate analysis.

The reason for the increased incidence of cardiovascular disease after treatment of testicular cancer is not clear. One suggestion is that chemotherapy causes vascular changes, which eventually translate into cardiovascular events. In one small study of 90 chemotherapy-treated testicular-cancer survivors, Nuver et al. showed that, compared to testicular-cancer patients who underwent surgery only, those who received chemotherapy (primarily BEP) had higher levels of markers of inflammation and endothelial activation, including fibrinogen, C-reactive protein, von Willebrand factor, plasminogen activator inhibitor, and tissue-type plasminogen activator (Nuver et al. 2004). Also, microalbuminuria, another reported predictor of cardiovascular events (Hillege et al. 2002), was seen in 10/90 of the chemotherapy patients (12 %) as compared to 1/44 of the surgery-only patients (2 %).

Although direct cardiac irradiation secondary to prophylactic mediastinal irradiation or by placing the superior edge of the retroperitoneal radiotherapy field within the thorax lead to cardiac toxicity, the etiology of cardiac toxicity following upper abdominal irradiation is less well understood and suggests an alternative mechanism. The renal vasculature and abdominal aorta receive full dose in standard seminoma treatment plans, suggesting perhaps a role for vascular endothelial damage. In a study of 147 head and neck and breast cancer patients undergoing reconstructive surgery following radiotherapy, intimal medial thickening was seen in arteries that had been irradiated (Russell et al. 2009). The authors postulated that radiation may cause inflammatory damage to the endothelial cells lining large vessels, leading to adhesion and transmigration of circulatory monocytes, which in the presence of cholesterol can become activated macrophages forming fatty streaks in the intima and thus initiate the process of atherosclerosis. While this explanation provides a plausible mechanism for the way in which excess cardiovascular disease might develop in the seminoma patient population who did not receive direct cardiac irradiation, irradiated testicular-cancer survivors received doses that were considerably lower than those used for other malignancies and have not been directly studied for vascular changes.

8.3   Decreased Fertility and Hypogonadism

In a study of 680 testicular-cancer survivors treated between 1982 and 1992, follicular stimulating hormone levels were increased following orchiectomy alone (41 %), radiotherapy (45 %), chemotherapy (49 %), and combined chemotherapy and radiation (71 %) (Nathanson et al. 2005), implying diminished spermatogenesis and risk of infertility. Fertility is approximately 30 % lower in testicular-cancer patients relative to the general population (Cvancarova et al. 2009) and is mainly dependent upon pretreatment gonadal function, age at diagnosis, and type of treatment, particularly chemotherapy (Fossa and Magelssen 2004, 2006; Kiserud et al. 2008; Cvancarova et al. 2009). In one study, among patients who received high-dose cisplatin-based chemotherapy (defined as greater than 850 mg), the 15-year posttreatment paternity rate was only 48 % (95 % CI = 30–69 %) (Brydoy et al. 2005). Radiotherapy can also affect spermatogenesis, and special

shielding of the contralateral testicle during irradiation is necessary to prevent permanent effects (Jacobsen et al. 1997). The paternity rate following the smaller PA-strip (as opposed to extended field) radiotherapy was 81 % ($p = .006$) (Brydoy et al. 2005). Although most testicular-cancer survivors who attempt paternity are successful (Brydoy et al. 2005), semen analysis and cryopreservation of sperm prior to cytotoxic therapy are recommended for most patients (Magelssen et al. 2005), as some will require assisted reproduction technology. Both are associated with increased rates of fertility in survivors (Syse 2008). Patients with dry ejaculation after receiving high-dose chemotherapy or non-nerve-sparing retroperitoneal lymph-node dissection are the least likely to achieve paternity (HR 0.19; 95 % CI, 0.10–0.34) (Brydoy et al. 2005).

Testosterone levels are generally normal after ipsilateral orchiectomy (Petersen et al. 1999) but may fall to the lower spectrum of normal after cisplatin-based chemotherapy (Brennemann et al. 1997; Huddart and Birtle 2005). It has been shown that 12–16 % of survivors are chemically hypogonadal (Nord et al. 2003; Huddart and Birtle 2005). The late effects of hypogonadism are not well studied in testicular-cancer survivors, although hypogonadism in other patient populations is known to be associated with metabolic syndrome, diabetes, cardiovascular disease, and osteoporosis (Corona et al. 2011a, b) In a large study of 823 testicular-cancer survivors followed for a median of 8 years, 15.2 % had chemical hypogonadism, 12.5 % had osteoporosis, and increased luteinizing hormone was seen in 15 % (Ondrusova et al. 2009).

Of interest, a statistically significant increase in the incidence of chronic fatigue, defined as lasting for more than 6 months, was reported among long-term Norwegian testicular-cancer survivors compared with age- and gender-matched referents (17 % vs. 10 %) (Orre et al. 2008). The authors found statistically significant higher levels of interleukin-1 and C-reactive protein in testicular-cancer survivors with fatigue than in those without fatigue (Orre et al. 2009). However, it is unknown if hypogonadism is also associated with fatigue. Impaired strength and muscular fatigue as "common complaints" among testicular-cancer patients who are receiving bleomycin, etoposide, and cisplatin-based chemotherapy is described by Christensen et al. (Christensen et al. 2011) although the exact proportion of patients with this concern and time course of muscular fatigue are not well studied at present. The authors plan a randomized trial evaluating the effect of strength on short-term training on muscular mass and muscular function in testicular-cancer patients undergoing chemotherapy, but there are no such studies in survivors.

8.4   Nephrotoxicity

The kidney is the principal excretory organ of cisplatin. The concentration of platinum in the kidney is several folds higher than in plasma and above any other organ (Launay-Vacher et al. 2008). As a result, cisplatin is a known nephrotoxin. The proximal nephron cells are believed to be most at risk, and damage here results in a decreased glomerular filtration rate.

With appropriate hydration and dosing, most exposed patients suffer an acute but reversible decrease in glomerular filtration rate; however, some experience irreversible damage, with up to a 30 % reduction in glomerular filtration (Hansen et al. 1988; Fossa et al. 2002). This is potentially very significant because even small changes in the glomerular filtration rate can be associated with both cardiovascular disease and all-cause mortality (Fossa et al. 2002; Astor et al. 2008). Platinum has been measured in the urine of long-term survivors following cisplatin-based chemotherapy (Gerl and Schierl 2000), which may provide a mechanism for ongoing renal damage.

The effect of radiotherapy on renal function has not been systematically studied. Depending upon the blocking techniques used and the position of the kidneys relative to the retroperitoneal nodes, which varies among individuals, the renal architecture in general receives very low doses of radiotherapy (see radiation figure). A very small study of 85 Norwegian testicular-cancer survivors included less than eight patients treated with radiotherapy alone for whom no change in creatinine was observed (Fossa et al. 2002). However, patients who received chemotherapy alone or

chemotherapy in combination with radiotherapy experienced a 20–30 % decline. There are, however, sporadic case reports of renal-artery stenosis following radiotherapy which can result in significant hypertension (Mulla et al. 2007). The risk factors for renal-artery stenosis are unknown, but stenosis is rare even following the higher therapeutic doses given for other cancer diagnoses (Salvi et al. 1983).

8.5   Neurotoxicity

Cisplatin accumulates in the dorsal nerve root ganglia where it can cause neuronal degeneration leading to sensory peripheral neuropathy (McKeage et al. 2001; McDonald et al. 2005). Patients may experience sensory loss, paresthesias, or Raynaud's phenomena, any of which can be permanent. One of the earliest studies of neurotoxicity looked at 30 Finnish survivors of metastatic germ cell tumor, all of whom had received high-dose cisplatin therapy (defined as six cycles) (Hansen et al. 1989). The interval between treatment and assessment was 4–9 years. Seventy-three percent had sensory loss and half complained of paresthesia. Nerve conduction studies demonstrated defects in 80 % of these survivors. A more recent cross-sectional study of 1,409 Norwegian testicular-cancer survivors revealed that with a median of follow-up time of 11 years, 39 % (95 % CI, 35–43 %) of men treated with chemotherapy reported Raynaud-like phenomena (defined as white or cold hands/fingers or feet/toes on cold exposure) and 29 % (95 % CI, 25–33 %) reported paresthesias in the hands or feet (Brydoy et al. 2009). The duration of neuropathy following chemotherapy is unclear from the available data, but there are no effective ameliorative treatments. Of note, men treated with radiotherapy had higher odds of self-reported paresthesias in their feet than those not treated with radiotherapy (OR = 1.5; 95 % CI, 1.01–2.1; $p = 0.04$). The latter finding was unexpected and has not been studied in other cohorts. In a separate study of 346 irradiated testicular-cancer survivors, transient neurological effects were seen in 11 (3.2 %). All but one patient experienced complete resolution of their symptoms by 1 year (Brydoy et al. 2007).

8.6 Ototoxicity

Because cisplatin has a low molecular weight, it crosses easily through the round window membrane into the cochlea (van der Hulst et al. 1988). The outer hair cells of the cochlea are believed to be most vulnerable. Patients treated with cisplatin display evidence of cochlear toxicity, high-frequency hearing loss (3–8 kHz), and tinnitus (Bauer and Brozoski 2005). In a mail survey of testicular-cancer survivors 35–44 years of age, the reported incidence of hearing loss was twice the national average for this age range (Stava et al. 2005). The prevalence of ototoxicity depends on the diligence of evaluation, as patients who report hearing loss may have other causes of hearing loss and also because some patients who report hearing loss have normal audiograms. In a study of 86 testicular-cancer survivors, 66 % had abnormal audiograms, but this percentage was decreased by excluding those whose abnormal audiograms were not typical of cisplatin damage (Bokemeyer et al. 1998). Subjectively, 17 (20 %) complained of persistent ototoxicity probably related to chemotherapy: tinnitus was experienced by ten (12 %), hearing loss by 3 (3 %) patients, and four (5 %) patients had both symptoms. A cross-sectional study of 1,319 Norwegian testicular-cancer survivors revealed that patients who received 1–4 cycles of cisplatin-based chemotherapy were more likely (OR=1.5; 95 % CI, 1.2–2.0) to experience hearing impairment and tinnitus compared with those who had not received chemotherapy (OR=1.8, 95 % CI, 1.4–2.4). However, the effect was most pronounced in those who received >5 cycles or dose-intensive chemotherapy; in these cohorts, the odds ratios ranged from 3.8 to 7.1 and were all highly statistically significant (Brydoy et al. 2009). Beyond the fact that it is bothersome, little is known about the impact of ototoxicity on quality of life or employment.

8.7 Genetic Susceptibility to Neurotoxicity and Ototoxicity

Along with the type and amount of cytotoxic drugs and radiotherapy, and comorbidities and habits, such as tobacco use (Travis et al. 2010),

genetic susceptibility may play a role in the development of late effects in testicular-cancer patients. Little is known about this although it is a topic of active study. There are data to suggest that common variants in TP53 and the gene for bleomycin hydrolase modify tumor response to chemotherapy and thus prognosis, but neither appears to predict for late effects, including pulmonary toxicity (Bergamaschi et al. 2003; de Haas et al. 2008; Maffei et al. 2008). One possible exception is the common codon 105 variant in glutathione S-transferase which detoxifies reactive metabolites of platinating agents and etoposide (Niitsu et al. 1998; Peklak-Scott et al. 2008). The substitution of isoleucine for valine at this site has been shown to be associated with increased risks of cisplatin-based oto- and neurotoxicity (Allan et al. 2001; Oldenburg et al. 2007; Oldenburg et al. 2008). Cisplatin can also be conjugated by other glutathione S-transferases, and there are early data to suggest that polymorphic functional variants may also be important in the subsequent development of ototoxicity (Peters et al. 2000; Oldenburg et al. 2007). Genetic testing of potentially important variants is not possible in the clinical setting at present but may be in the future.

8.8 Pulmonary Toxicity

Pulmonary function has been shown in population studies to be a predictor of both all-cause mortality and quality of life (Schunemann et al. 2000; Mannino et al. 2003). In an international population-based registry study of more than 38,000 testicular-cancer survivors, an increased risk of death from respiratory disease was shown with a standardized mortality ratio of 1.15 (95 % CI, 0.99–1.34) (Fossa et al. 2007). Patients who received chemotherapy during the cisplatin–bleomycin era (after 1975) had threefold more excess deaths from pulmonary causes. Bleomycin exerts its antitumor effect by inducing programmed cell death, inhibiting angiogenesis, and by the formation of free radicals. Bleomycin is excreted through the kidneys and deactivated by the enzyme bleomycin hydrolase. The lung contains no bleomycin hydrolase and, as a result, is particularly vulnerable to bleomycin toxicity.

Short-term pulmonary complications of bleomy-cin are seen in up to 46 % of patients but most are mild and self-limited (Sleijfer 2001). The risk factors for bleomycin-associated pneumonitis are well described elsewhere, and only a small fraction of bleomycin-treated patients develop pulmonary fibrosis. The risk of fatality in patients who develop bleomycin-associated pulmonary fibrosis is on the order of 10 % (Dearnaley et al. 1991; Sleijfer 2001; O'Sullivan et al. 2003). Cisplatin is not generally thought to be associated with pulmonary late effects. However, a recent study of late effects in 1,000 chemotherapy-treated testicular-cancer survivors followed for a median of 11.2 years revealed that age-adjusted forced vital capacity (FVC), FVC predicted, forced expiratory volume 1 (FEV1), and FEV1 predicted were all significantly lower in patients treated with higher-dose cisplatin (<850 mg) or cisplatin with pulmonary surgery compared with patients who only underwent surgery (Haugnes et al. 2009). Eight percent of all study patients had restrictive lung disease, and the highest prevalence was in the higher-dose group (17.7 %) and in the cisplatin with pulmonary surgery group (16.7 %). Compared with patients who underwent surgery only, these groups had odds ratio for restrictive disease of 3.1 (95 % CI, 1.3–7.3) and 2.5 (95 % CI, 0.8–7.6), respectively. In that study, only cisplatin dose ($p=0.007$) and age at diagnosis ($p=0.008$) were associated with restrictive lung disease even when bleomycin was included in the model. To date, other than exposure to chemotherapy and ongoing tobacco use, there are no prediction models which can be used to identify survivors at risk for either pulmonary fibrosis or death from pulmonary fibrosis.

8.9 Late Psychosocial Effects

8.9.1 Sexuality and Marriage

A recent diagnosis of testicular cancer was associated with a 20 % increased probability of divorce over the first 5 years (OR=1.19; 95 % CI, 1.04–1.38) although the effect appeared to level off after that (Syse 2008). Little is known about the interplay among infertility, divorce, and hypogo-nadism for testicular-cancer patients in paired relationships. One small study reported that couples whose relationships began before a diagnosis of testicular cancer were more sexually satisfied than those whose relationships began after (Tuinman et al. 2005). Patients who underwent non-nerve-sparing retroperitoneal lymph-node dissection, as was popular for stage I or II disease until the mid-to-late 1980s (Lange et al. 1987), suffered from sympathetic nerve dysfunction leading to retrograde ejaculation and infertility. Infertility-related effects on relationships may be important but have not been studied since 1990, when a series of three publications on 200 men with testicular cancer revealed that distress over infertility was associated most with childlessness and that fertility was important to men with testicular cancer (Rieker et al. 1985, 1989, 1990).

A recent study of sexual dysfunction among Norwegian testicular-cancer survivors demonstrated a 39 % incidence of sexual dysfunction compared with 36 % in referents. However, the testicular-cancer survivors had significantly worse scores pertaining to ejaculatory and sexual problems (Dahl et al. 2007). Overall, sexual problems were significantly associated with increasing age, lack of a partner, and higher anxiety scores. In that study, survivors reported slightly better rates of sexual satisfaction when compared to the normative sample; the authors attributed this, in part, to response shift (Schwartz et al. 2007), which represents adaptation of attitude and expectations in response to disease and treatment.

8.9.2 Mental Health and Quality of Life

In a large study of 1,400 testicular-cancer survivors, mental health and quality of life were assessed by validated questionnaires; no significant differences were seen between survivors and referents (Dahl et al. 2005b), similar to other reports (Heidenreich and Hofmann 1999). Testicular-cancer patients who reported more side effects and particularly those who reported stress related to testicular cancer (i.e., fear of recurrence) were more likely to score lower in terms of quality of life. It is not clear, however, that the questionnaires addressed concerns specific to testicular-cancer patients (Travis et al. 2010). In one report looking specifically at anxiety and depression scores in 1,400 testicular-cancer survivors using the Hospital Anxiety and Depression Scale (HADS), HADS-defined anxiety disorder was more prevalent in testicular-cancer survivors than in the normative sample (RR = 1.49; 95 % CI, 1.31–1.69) although the incidence of HADS-defined depression did not differ from the norm (Dahl et al. 2005a). In this study, HADS-defined anxiety disorder was associated on multivariate analysis with young age, peripheral neuropathy, economic problems, alcohol problems, relapse anxiety, and prior treatment for psychological problems. The authors attributed the increased incidence of HADS-defined anxiety disorder in part to the surprisingly high proportion of survivors (31 %) who reported relapse anxiety even after a median follow-up time of almost 11 years. Relapse anxiety and HADS-defined anxiety disorder were highly correlated in this study. In point of fact, a review of late relapse, defined as relapse more than 2 years following successful treatment for testicular cancer, showed that the actual risk is only 1–4 % (Oldenburg et al. 2006). In a separate study, hypogonadal patients, representing 13 % of 680 testicular-cancer survivors, had lower quality-of-life scores related to sexual functioning and lower physical, social, and role function scores on the European Organization for Research and the Treatment of Cancer Quality-of-Life form C-30 (Huddart et al. 2005).

8.9.3 Employment

A meta-analysis of 26 articles describing 36 studies of employment in cancer survivors included 20,366 survivors and 157,603 healthy control participants (de Boer et al. 2009). This analysis included studies from the United States, Europe, and other countries. Overall, cancer survivors were more likely to be unemployed than the healthy controls (33.8 % vs. 15.2 %; RR = 1.37; 95 % CI, 1.21–1.55). However, the risk of unemployment was not higher for survivors of testicular cancer

(18.5 % vs. 18.1 %; pooled RR = 0.94; 95 % CI, 0.74–1.20).

8.10 Summary

In summary, the majority of men with testicular cancer are cured of their malignancy and survive for decades. Over time, however, they develop late effects of treatment or continue to experience side effects that developed during therapy and persist. The most serious late effects are second malignant neoplasms and cardiovascular disease as these are potentially life threatening. Cardiovascular precursors, such as metabolic syndrome, hypertension, and diabetes, are present in testicular-cancer survivors as well. Fatigue, anxiety, infertility, hypogonadism, ototoxicity, neurotoxicity, and pulmonary toxicity can all potentially affect quality of life. This patient population has not been tested sufficiently for development of evidence-based guidelines for their continuing care, but efforts are underway to develop consensus-based approaches for their management. Clinicians caring for these patients should encourage them to follow the generally available guidelines for good health such as weight management, regular exercise, and avoidance of known carcinogens, such as tobacco. Probably the best option for the future good health for testicular-cancer patients is the avoidance of unnecessary exposure to chemotherapy and radiation, and present-day treatment programs are designed to decrease treatment intensity while maintaining very high cure rates.

References

Allan JM, Wild CP, Rollinson S et al (2001) Polymorphism in glutathione S-transferase P1 is associated with susceptibility to chemotherapy-induced leukemia. Proc Natl Acad Sci U S A 98:11592–11597
Astor BC, Hallan SI, Miller ER 3rd et al (2008) Glomerular filtration rate, albuminuria, and risk of cardiovascular and all-cause mortality in the US population. Am J Epidemiol 167:1226–1234
Bauer CA, Brozoski TJ (2005) Cochlear structure and function after round window application of ototoxins. Hear Res 201:121–131

Bergamaschi D, Gasco M, Hiller L et al (2003) p53 polymorphism influences response in cancer chemotherapy via modulation of p73-dependent apoptosis. Cancer Cell 3:387–402

Bokemeyer C, Berger CC, Hartmann JT et al (1998) Analysis of risk factors for cisplatin-induced ototoxicity in patients with testicular cancer. Br J Cancer 77:1355–1362

Bokemeyer C, Schmoll HJ (1993) Secondary neoplasms following treatment of malignant germ cell tumors. J Clin Oncol 11:1703–1709

Brennemann W, Stoffel-Wagner B, Helmers A et al (1997) Gonadal function of patients treated with cisplatin based chemotherapy for germ cell cancer. J Urol 158:844–850

Brydoy M, Fossa SD, Klepp O et al (2005) Paternity following treatment for testicular cancer. J Natl Cancer Inst 97:1580–1588

Brydoy M, Oldenburg J, Klepp O et al (2009) Observational study of prevalence of long-term Raynaud-like phenomena and neurological side effects in testicular cancer survivors. J Natl Cancer Inst 101:1682–1695

Brydoy M, Storstein A, Dahl O (2007) Transient neurological adverse effects following low dose radiation therapy for early stage testicular seminoma. Radiother Oncol 82:137–144

Christensen JF, Andersen JL, Adamsen L et al (2011) Progressive resistance training and cancer testis (PROTRACT) – efficacy of resistance training on muscle function, morphology and inflammatory profile in testicular cancer patients undergoing chemotherapy: design of a randomized controlled trial. BMC Cancer 11:326

Corona G, Rastrelli G, Monami M et al (2011a) Hypogonadism as a risk factor for cardiovascular mortality in men: a meta-analytic study. Eur J Endocrinol 165:687–701

Corona G, Rastrelli G, Vignozzi L et al (2011b) Testosterone, cardiovascular disease and the metabolic syndrome. Best Pract Res Clin Endocrinol Metab 25:337–353

Cvancarova M, Samuelsen SO, Magelssen H et al (2009) Reproduction rates after cancer treatment: experience from the Norwegian radium hospital. J Clin Oncol 27:334–343

Dahl AA, Bremnes R, Dahl O et al (2007) Is the sexual function compromised in long-term testicular cancer survivors? Eur Urol 52:1438–1447

Dahl AA, Haaland CF, Mykletun A et al (2005a) Study of anxiety disorder and depression in long-term survivors of testicular cancer. J Clin Oncol 23:2389–2395

Dahl AA, Mykletun A, Fossa SD (2005b) Quality of life in survivors of testicular cancer. Urol Oncol 23:193–200

de Boer AG, Taskila T, Ojajarvi A et al (2009) Cancer survivors and unemployment: a meta-analysis and meta-regression. JAMA 301:753–762

de Haas EC, Zwart N, Meijer C et al (2008) Variation in bleomycin hydrolase gene is associated with reduced

survival after chemotherapy for testicular germ cell cancer. J Clin Oncol 26:1817–1823

Dearnaley DP, Horwich A, A'Hern R et al (1991) Combination chemotherapy with bleomycin, etoposide and cisplatin (BEP) for metastatic testicular teratoma: long-term follow-up. Eur J Cancer 27:684–691

Fossa SD, Aass N, Winderen M et al (2002) Long-term renal function after treatment for malignant germ-cell tumours. Ann Oncol 13:222–228

Fossa SD, Chen J, Schonfeld SJ et al (2005) Risk of contralateral testicular cancer: a population-based study of 29,515 U.S. men. J Natl Cancer Inst 97:1056–1066

Fossa SD, Gilbert E, Dores GM et al (2007) Noncancer causes of death in survivors of testicular cancer. J Natl Cancer Inst 99:533–544

Fossa SD, Magelssen H (2004) Fertility and reproduction after chemotherapy of adult cancer patients: malignant lymphoma and testicular cancer. Ann Oncol 15(Suppl 4):iv259–iv265

Gerl A, Schierl R (2000) Urinary excretion of platinum in chemotherapy-treated long-term survivors of testicular cancer. Acta Oncol 39:519–522

Hansen SW, Groth S, Daugaard G et al (1988) Long-term effects on renal function and blood pressure of treatment with cisplatin, vinblastine, and bleomycin in patients with germ cell cancer. J Clin Oncol 6:1728–1731

Hansen SW, Helweg-Larsen S, Trojaborg W (1989) Long-term neurotoxicity in patients treated with cisplatin, vinblastine, and bleomycin for metastatic germ cell cancer. J Clin Oncol 7:1457–1461

Haugnes HS, Aass N, Fossa SD et al (2009) Pulmonary function in long-term survivors of testicular cancer. J Clin Oncol 27:2779–2786

Haugnes HS, Wethal T, Aass N et al (2010) Cardiovascular risk factors and morbidity in long-term survivors of testicular cancer: a 20-year follow-up study. J Clin Oncol 28(30):4649–4657

Heidenreich A, Hofmann R (1999) Quality-of-life issues in the treatment of testicular cancer. World J Urol 17:230–238

Hemminki K, Liu H, Sundquist J (2010) Second cancers after testicular cancer diagnosed after 1980 in Sweden. Ann Oncol 21:1546–1551

Hillege HL, Fidler V, Diercks GF et al (2002) Urinary albumin excretion predicts cardiovascular and noncardiovascular mortality in general population. Circulation 106:1777–1782

Huddart RA, Norman A, Shahidi M et al (2003) Cardiovascular disease as a long-term complication of treatment for testicular cancer. J Clin Oncol 21(8):1513–1523

Huddart RA, Birtle AJ (2005) Recent advances in the treatment of testicular cancer. Expert Rev Anticancer Ther 5:123–138

Huddart RA, Norman A, Moynihan C et al (2005) Fertility, gonadal and sexual function in survivors of testicular cancer. Br J Cancer 93:200–207

Jacobsen KD, Olsen DR, Fossa K et al (1997) External beam abdominal radiotherapy in patients with seminoma stage I: field type, testicular dose, and sper-matogenesis. Int J Radiat Oncol Biol Phys 38:95–102

Kiserud CE, Magelssen H, Fedorcsak P et al (2008) Gonadal function after cancer treatment in adult men. Tidsskr Nor Laegeforen 128:461–465

Kollmannsberger C, Hartmann JT, Kanz L et al (1999) Therapy-related malignancies following treatment of germ cell cancer. Int J Cancer 83:860–863

Lange PH, Chang WY, Fraley EE (1987) Fertility issues in the therapy of nonseminomatous testicular tumors. Urol Clin North Am 14:731–747

Launay-Vacher V, Rey JB, Isnard-Bagnis C et al (2008) Prevention of cisplatin nephrotoxicity: state of the art and recommendations from the European Society of Clinical Pharmacy Special Interest Group on Cancer Care. Cancer Chemother Pharmacol 61:903–909

Maffei F, Carbone F, Angelini S et al (2008) Micronuclei frequency induced by bleomycin in human peripheral lymphocytes: correlating BLHX polymorphism with mutagen sensitivity. Mutat Res 639:20–26

Magelssen H, Brydoy M, Fossa SD (2006) The effects of cancer and cancer treatments on male reproductive function. Nat Clin Pract Urol 3:312–322

Magelssen H, Haugen TB, von During V et al (2005) Twenty years experience with semen cryopreservation in testicular cancer patients: who needs it? Eur Urol 48:779–785

Mannino DM, Buist AS, Petty TL et al (2003) Lung function and mortality in the United States: data from the First National Health and Nutrition Examination Survey follow up study. Thorax 58:388–393

McDonald ES, Randon KR, Knight A et al (2005) Cisplatin preferentially binds to DNA in dorsal root ganglion neurons in vitro and in vivo: a potential mechanism for neurotoxicity. Neurobiol Dis 18:305–313

McKeage MJ, Hsu T, Screnci D et al (2001) Nucleolar damage correlates with neurotoxicity induced by different platinum drugs. Br J Cancer 85:1219–1225

Mulla MG, Ananthkrishnan G, Mirza MS (2007) Renal artery stenosis after radiotherapy for stage I seminoma, a case report. Clin Oncol (R Coll Radiol) 19:209

Nathanson KL, Kanetsky PA, Hawes R et al (2005) The Y deletion gr/gr and susceptibility to testicular germ cell tumor. Am J Hum Genet 77:1034–1043

Niitsu Y, Takahashi Y, Ban N et al (1998) A proof of glutathione S-transferase-pi-related multidrug resistance by transfer of antisense gene to cancer cells and sense gene to bone marrow stem cell. Chem Biol Interact 111–112:325–332

Nord C, Bjoro T, Ellingsen D et al (2003) Gonadal hormones in long-term survivors 10 years after treatment for unilateral testicular cancer. Eur Urol 44:322–328

Nuver J, Smit AJ, Sleijfer DT et al (2004) Microalbuminuria, decreased fibrinolysis, and inflammation as early signs of atherosclerosis in long-term survivors of disseminated testicular cancer. Eur J Cancer 40:701–706

O'Sullivan JM, Huddart RA, Norman AR et al (2003) Predicting the risk of bleomycin lung toxicity in patients with germ-cell tumours. Ann Oncol 14:91–96

Oldenburg J, Fossa SD, Ikdahl T (2008) Genetic variants associated with cisplatin-induced ototoxicity. Pharmacogenomics 9:1521–1530

Oldenburg J, Kraggerud SM, Brydoy M et al (2007) Association between long-term neuro-toxicities in testicular cancer survivors and polymorphisms in glutathione-s-transferase-P1 and -M1, a retrospective cross sectional study. J Transl Med 5:70

Oldenburg J, Martin JM, Fossa SD (2006) Late relapses of germ cell malignancies: incidence, management, and prognosis. J Clin Oncol 24:5503–5511

Ondrusova M, Ondrus D, Dusek L et al (2009) Damage of hormonal function and bone metabolism in long-term survivors of testicular cancer. Neoplasma 56:473–479

Orre IJ, Fossa SD, Murison R et al (2008) Chronic cancer-related fatigue in long-term survivors of testicular cancer. J Psychosom Res 64:363–371

Orre IJ, Murison R, Dahl AA et al (2009) Levels of circulating interleukin-1 receptor antagonist and C-reactive protein in long-term survivors of testicular cancer with chronic cancer-related fatigue. Brain Behav Immun 23:868–874

Osterlind A, Berthelsen JG, Abildgaard N et al (1991) Risk of bilateral testicular germ cell cancer in Denmark: 1960–1984. J Natl Cancer Inst 83:1391–1395

Pedersen-Bjergaard J, Daugaard G, Hansen SW et al (1991) Increased risk of myelodysplasia and leukaemia after etoposide, cisplatin, and bleomycin for germ-cell tumours. Lancet 338:359–363

Peklak-Scott C, Smitherman PK, Townsend AJ et al (2008) Role of glutathione S-transferase P1-1 in the cellular detoxification of cisplatin. Mol Cancer Ther 7:3247–3255

Peters U, Preisler-Adams S, Hebeisen A et al (2000) Glutathione S-transferase genetic polymorphisms and individual sensitivity to the ototoxic effect of cisplatin. Anticancer Drugs 11:639–643

Petersen PM, Skakkebaek NE, Rorth M et al (1999) Semen quality and reproductive hormones before and after orchiectomy in men with testicular cancer. J Urol 161:822–826

Richiardi L, Mirabelli D, Calisti R et al (2006) Occupational exposure to diesel exhausts and risk for lung cancer in a population-based case-control study in Italy. Ann Oncol 17(12):1842–1847

Richiardi L, ScÈlo G, Boffetta P et al (2007) Second malignancies among survivors of germ-cell testicular cancer: a pooled analysis between 13 cancer registries. Int J Cancer 120:623–631

Rieker PP, Edbril SD, Garnick MB (1985) Curative testis cancer therapy: psychosocial sequelae. J Clin Oncol 3:1117–1126

Rieker PP, Fitzgerald EM, Kalish LA (1990) Adaptive behavioral responses to potential infertility among survivors of testis cancer. J Clin Oncol 8:347–355

Rieker PP, Fitzgerald EM, Kalish LA et al (1989) Psychosocial factors, curative therapies, and behavioral outcomes. A comparison of testis cancer survivors and a control group of healthy men. Cancer 64:2399–2407

Russell NS, Hoving S, Heeneman S et al (2009) Novel insights into pathological changes in muscular arteries of radiotherapy patients. Radiother Oncol 92:477–483

Salvi S, Green DM, Brecher ML et al (1983) Renal artery stenosis and hypertension after abdominal irradiation for Hodgkin disease. Successful treatment with nephrectomy. Urology 21:611–615

Schunemann HJ, Dorn J, Grant BJ et al (2000) Pulmonary function is a long-term predictor of mortality in the general population: 29-year follow-up of the Buffalo Health Study. Chest 118:656–664

Schwartz CE, Andresen EM, Nosek MA et al (2007) Response shift theory: important implications for measuring quality of life in people with disability. Arch Phys Med Rehabil 88:529–536

Sleijfer S (2001) Bleomycin-induced pneumonitis. Chest 120:617–624

Stava C, Beck M, Schultz PN et al (2005) Hearing loss among cancer survivors. Oncol Rep 13:1193–1199

Syse A (2008) Does cancer affect marriage rates? J Cancer Surviv 2:205–214

Travis LB, Weeks J, Curtis RE et al (1996) Leukemia following low-dose total body irradiation and chemotherapy for non-Hodgkin's lymphoma. J Clin Oncol 14:565–571

Travis LB, Andersson M, Gospodarowicz M et al (2000) Treatment-associated leukemia following testicular cancer. J Natl Cancer Inst 92:1165–1171

Travis LB, Fossa SD, Schonfeld SJ et al (2005) Second cancers among 40,576 testicular cancer patients: focus on long-term survivors. J Natl Cancer Inst 97:1354–1365

Travis LB, Beard C, Allan JM et al (2010) Testicular cancer survivorship: research strategies and recommendations. J Natl Cancer Inst 102(15):1114–1130

Travis LB, Beard C, Allan JM et al (2011) Testicular cancer survivorship: research strategies and recommendations. J Natl Cancer Inst 102(15):1114–1130

Tuinman MA, Fleer J, Sleijfer DT et al (2005) Marital and sexual satisfaction in testicular cancer survivors and their spouses. Support Care Cancer 13:540–548

van den Belt-Dusebout AW, de Wit R, Gietema JA et al (2007) Treatment-specific risks of second malignancies and cardiovascular disease in 5-year survivors of testicular cancer. J Clin Oncol 25:4370–4378

van den Belt-Dusebout AW, Nuver J, de Wit R et al (2006) Long-term risk of cardiovascular disease in 5-year survivors of testicular cancer. J Clin Oncol 24:467–475

van der Hulst RJ, Dreschler WA, Urbanus NA (1988) High frequency audiometry in prospective clinical research of ototoxicity due to platinum derivatives. Ann Otol Rhinol Laryngol 97:133–137

van Leeuwen FE, Stiggelbout AM, van den Belt-Dusebout AW et al (1993) Second cancer risk following testicular cancer: a follow-up study of 1,909 patients. J Clin Oncol 11:415–424

Wanderas EH, Fossa SD, Tretli S (1997) Risk of subsequent non-germ cell cancer after treatment of germ cell cancer in 2006 Norwegian male patients. Eur J Cancer 33(2):253–262

Wethal T, Kjekshus J, Roislien J et al (2007) Treatment-related differences in cardiovascular risk factors in long-term survivors of testicular cancer. J Cancer Surviv 1:8–16

Zagars GK, Ballo MT, Lee AK et al (2004) Mortality after cure of testicular seminoma. J Clin Oncol 22:640–647

Index